AF870962

Updates on Supercapacitors

Edited by Zoran Stevic

Published in London, United Kingdom

Updates on Supercapacitors
http://dx.doi.org/10.5772/intechopen.100764
Edited by Zoran Stevic

Contributors
Kefayat Ullah, Bakht Mand Khan, Amin Ur Rashid, Won Chun Oh, Mohamed M. Atta, Rania M. Ahmed, Zoran M. Stevic, Ilija Radovanovic, Yi Su, Mohamad Sawan, Arun Aravind, Nithya S. George, Lolly Maria Jose

© The Editor(s) and the Author(s) 2023
The rights of the editor(s) and the author(s) have been asserted in accordance with the Copyright, Designs and Patents Act 1988. All rights to the book as a whole are reserved by INTECHOPEN LIMITED. The book as a whole (compilation) cannot be reproduced, distributed or used for commercial or non-commercial purposes without INTECHOPEN LIMITED's written permission. Enquiries concerning the use of the book should be directed to INTECHOPEN LIMITED rights and permissions department (permissions@intechopen.com).
Violations are liable to prosecution under the governing Copyright Law.

Individual chapters of this publication are distributed under the terms of the Creative Commons Attribution 3.0 Unported License which permits commercial use, distribution and reproduction of the individual chapters, provided the original author(s) and source publication are appropriately acknowledged. If so indicated, certain images may not be included under the Creative Commons license. In such cases users will need to obtain permission from the license holder to reproduce the material. More details and guidelines concerning content reuse and adaptation can be found at http://www.intechopen.com/copyright-policy.html.

Notice
Statements and opinions expressed in the chapters are these of the individual contributors and not necessarily those of the editors or publisher. No responsibility is accepted for the accuracy of information contained in the published chapters. The publisher assumes no responsibility for any damage or injury to persons or property arising out of the use of any materials, instructions, methods or ideas contained in the book.

First published in London, United Kingdom, 2023 by IntechOpen
IntechOpen is the global imprint of INTECHOPEN LIMITED, registered in England and Wales, registration number: 11086078, 5 Princes Gate Court, London, SW7 2QJ, United Kingdom

British Library Cataloguing-in-Publication Data
A catalogue record for this book is available from the British Library

Additional hard and PDF copies can be obtained from orders@intechopen.com

Updates on Supercapacitors
Edited by Zoran Stevic
p. cm.
Print ISBN 978-1-83962-641-8
Online ISBN 978-1-83962-642-5
eBook (PDF) ISBN 978-1-83962-688-3

We are IntechOpen, the world's leading publisher of Open Access books

Built by scientists, for scientists

6,200+
Open access books available

168,000+
International authors and editors

185M+
Downloads

156
Countries delivered to

Our authors are among the
Top 1%
most cited scientists

12.2%
Contributors from top 500 universities

CLARIVATE ANALYTICS
BOOK CITATION INDEX
INDEXED

WEB OF SCIENCE™

Selection of our books indexed in the Book Citation Index
in Web of Science™ Core Collection (BKCI)

Interested in publishing with us?
Contact book.department@intechopen.com

Numbers displayed above are based on latest data collected.
For more information visit www.intechopen.com

Meet the editor

Zoran M. Stevic is a full professor at the University of Belgrade, Technical Faculty in Bor and the School of Electrical Engineering in Belgrade, Serbia. He received his Ph.D. from the School of Electrical Engineering at the University of Belgrade. His research areas include power electronics, IR thermography, computer measurement and control, supercapacitors, energy efficiency, renewable energy sources, system modeling, sensors and optoelectronics. He is a member of IEEE, IBPSA, SMEITS and SCS. He has been a project coordinator and a member of working groups on many scientific and technical projects. He has published over 350 papers (including 60 fully refereed papers in scientific journals), five books and six chapters. He is a thermography professional.

Contents

Preface

For decades, science has been intensively researching electrochemical systems that exhibit extremely high, and previously unattainable, capacitance values (on the order of hundreds of Fg^{-1}). Research has not only demonstrated the unsuspected possibilities of supercapacitors, but also marked out a new direction for the development of electrical energy storage systems [1]. With the recent development of new materials and technologies, very large developed surfaces and very small inter-electrode distances have been achieved. Enormous pseudo-capacitance of several orders of magnitude larger than standard capacitors has been achieved for many materials, and such systems are called supercapacitors or, less frequently, ultracapacitors) [2].

There are two types of supercapacitors, depending on the energy storage mechanism. Electric double-layer capacitors use the electrostatic principle, and for pseudo-capacitors, [3] the charge storage is caused by fast redox reactions [4]. Electrode systems using both mechanisms are called hybrid supercapacitors. Supercapacitors are widely used due to their fast charge and discharge, and a huge number of charge/discharge cycles [5].

Chapter 1 focuses on the need for further development and improvement of supercapacitor properties to power isolated systems (sensor networks, IoT), covering peaks of electricity consumption, filtration and new, more efficient topologies in power electronics. New technologies and materials, and their fabrication, modeling, characterization and applications are also presented.

Carbon allotropes, including fullerenes, carbon nanotubes, and graphene-based supercapacitors, are discussed in Chapter 2, while graphene-based nanocomposites for supercapacitor electrodes are presented in Chapter 3.

Transition metal oxides (TMOs) are pseudo-capacitor electrode materials that show high specific capacitance and are a better redox active material for energy storage applications [6]. Chapter 4 examines developments in the fabrication of six TMO-based electrode materials (NiO, ZnO, MnO2, SnO_2, WO_3, V_2O_5) for enhanced electrochemical performance.

In Chapter 5, the authors show how various technologies are used to fabricate electrodes and supercapacitors, and present several applications.

Zoran Stevic
Technical Faculty in Bor,
School of Electrical Engineering,
University of Belgrade,
Belgrade, Serbia

References

[1] Conway BE. Electrochemical Supercapacitors. New York: Kluwer Academic/Plenum Publishers; 1999

[2] Stevic Z. Very large capacitance capacitors based on copper sulfide [thesis]. School of Electrical Engineering, University of Belgrade; 2004

[3] Xiong S, Jiang S, Wang J, Lin H, Lin M, Weng S, et al. A high-performance hybrid supercapacitor with NiO derived NiO@Ni-MOF composite electrodes. Electrochimica Acta. 2020. DOI: 10.1016/j.electacta.2020.135956

[4] Adib K, Sohouli E. Sonochemical synthesis of Ag2WO4/RGO-based nanocomposite as a potential material for supercapacitors electrodes. Ceramics International. 2021;**47**(10):14075-14086

[5] Park J, Lee J, Kim S, Hwang J. Graphene-based two-dimensional mesoporous materials: Synthesis and electrochemical energy storage applications. Materials. 2021;**14**(10):2597. DOI: 10.3390/ma14102597

[6] Liang R, et al. Transition metal oxide electrode materials for supercapacitors: A review of recent developments. Nanomaterials. 2021;**11**(5):1248

Chapter 1

Supercapacitors: The Innovation of Energy Storage

Zoran Stevic and Ilija Radovanovic

Abstract

In addition to the accelerated development of standard and novel types of rechargeable batteries, for electricity storage purposes, more and more attention has recently been paid to supercapacitors as a qualitatively new type of capacitor. A large number of teams and laboratories around the world are working on the development of supercapacitors, while their ever-improving performances enable wider use. The major challenges are to improve the parameters of supercapacitors, primarily energy density and operating voltage, as well as the miniaturization, optimization, energy efficiency, economy, and environmental acceptance. This chapter provides an overview of new techniques and technologies of supercapacitors that are changing the present and future of electricity storage, with special emphasis on self-powering sensor and transmitter systems. The latest achievements in the production, modeling, and characterization of supercapacitor elements (electrode materials, electrolytes, and supporting elements) whose parameters are optimized for long-term self-supply of low power consumers (low voltage, high energy density, and low leakage current, etc.) are considered.

Keywords: supercapacitors, innovation, energy storage, application

1. Introduction

For decades, science has been intensively researching electrochemical systems that exhibit extremely high capacitance values (in the order of hundreds of Fg^{-1}), which were previously unattainable. The early researches have shown the unsuspected possibilities of supercapacitors and traced a new direction for the development of electrical energy storage systems [1]. In recent times, with the development of new materials and technologies, very large developed surfaces and very small inter-electrode distances have been achieved. In many materials, enormous pseudocapacitance is also expressed, which achieves extremely large capacitances (several orders of magnitude larger than standard capacitors), so such systems are called supercapacitors (supercapacitors or, more rarely, ultracapacitors) [2].

There are two types of supercapacitors, depending on the energy storage mechanism: electric double-layer capacitors and pseudocapacitors [3]. In the first case, it is an electrostatic principle, and in the second one, the charge storage is caused by fast redox reactions [4]. Some electrode materials have both one and the other mechanism, thus so-called hybrid capacitors are formed on their basis.

High needs for powering isolated systems (sensor networks and IoT), covering peaks of electricity consumption, filtration, and new more efficient topologies in power electronics require further development and improvement of supercapacitor properties. More on this topic is given in Section 2 of this chapter.

New technologies and new materials in this area are presented in Section 3 of this chapter. Fabrication, modeling, and characterization are presented in Section 4.

Supercapacitors are widely used due to their high-power density, that is, fast charge and discharge, and a huge number of charge/discharge cycles [5]. Increasing the performance of supercapacitors opens up new fields of application and attention will be paid to this in Section 5. The growth of the industry in this area causes a drop in prices, which will be discussed in Section 6.

2. Need for supercapacitors

Since the energy harvesting from renewable energy sources is highly actual today, the studies are also focused on the diverse methods for storing this energy in the form of electricity. Supercapacitors are one of the most efficient energy storage devices. As they have many advantages, supercapacitors are continuously being used in devices and systems that are eager for a high-power supply, opposite to the batteries. Recently actual, supercapacitors' applications are driven by their high performance and market potential, placing them in many fields of interest, such as industrial control, power, transportation, consumer electronic products, national defense, communications, medical equipment, electric and hybrid vehicles [6–8].

Nowadays, with the rapid development of intelligent electronic devices, have placed flexible energy storage devices in the focus of researchers. The industry requires energy storage that are flexible and optimized but endowed with high electrochemical properties [8–10]. The advantages of the supercapacitors, such as charge-discharge cycle life, size and weight, and environmentally oriented, suiting them for various applications. Supercapacitors are being used more and more as applications require storing and releasing high amounts of energy in short periods. Current industry applications include the automotive industry, hybrid transportation systems around the world, grid stabilization, utility vehicles, and rail-system power models [11].

The storing of energy is one of the main applications of supercapacitors. Following their outstanding power characteristics, supercapacitors are vital for the energy sector and their stationary applications. Additionally, the low maintenance requirements, as well as the extreme conditions that supercapacitors are able to withstand, make them suitable for renewable energy-related applications [12, 13]. Furthermore, the supercapacitors provide substantial benefits to railway electricity systems and the aerospace industry, since these sectors are trying to achieve a more electric power supply [13, 14]. Furthermore, many systems in the industrial sector are using supercapacitors, including small vehicles, such as forklifts, shovel trucks, agricultural machinery, excavators, mining shovels, harbor cranes, and industrial lasers. Consumer electronics are relying on supercapacitors, especially in real-time clock or memory backup, power failure backup, storage applications in which supercapacitors are used instead of batteries, and high load assistance to the primary electrical energy storage systems [13].

3. New technologies and materials for supercapacitors

Supercapacitors are increasingly used for energy storage due to their large number of charge and discharge cycles, high power density, minimal maintenance, long life span, and environmental friendliness [15]. The only disadvantage over batteries, the lower energy density, is decreasing more and more thanks to the intensive development of new technologies and new materials. Hybrid electrodes that combine double-layer (electrostatic) capacitance and pseudo (redox) capacitance are increasingly being used [16].

The latest nanotechnologies have given rise to nanomaterials, such as 2D graphene, 1D CNT, and 0D fullerene [17]. There is a high trend in the research of carbon-based electrode materials (CNT, graphene, fullerene and their composites with metal oxides, etc.), copper sulfide and other metals, and metal oxides, all in combination with appropriate electrolytes.

For years, the authors of this chapter have researched the possibility of using natural copper minerals, primarily chalcosine [18] and coveline [19].

The latest research in the field of perovskite oxide applications for supercapacitor electrode materials deserves special attention. Perovskite oxides based on lanthanum, strontium, and cerium, etc. are being researched [20].

The researches with polymer materials are of great interest as well, which, in addition to high power density, also has an acceptable energy density (**Figure 1**). Conductive polymers are materials that contain a conjugated double bond, which places them in the group of materials that exhibit good electrical conductivity. Apart from showing electrical conductivity, these materials are characterized by redox reactions that take place when the electrode is polarized in a certain potential range. As a consequence of the redox reaction, constant currents are recorded in a wide range of potentials, which indicates the continuous development of the redox reaction in the

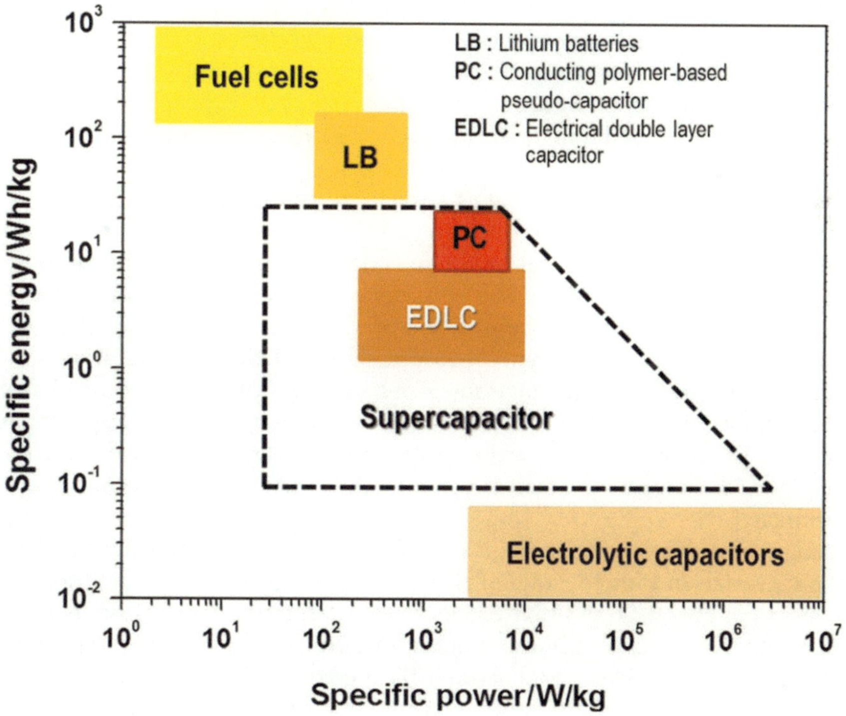

Figure 1.
Ragone plot shows the energy vs. power density comparison of supercapacitors with the other energy storage devices [21].

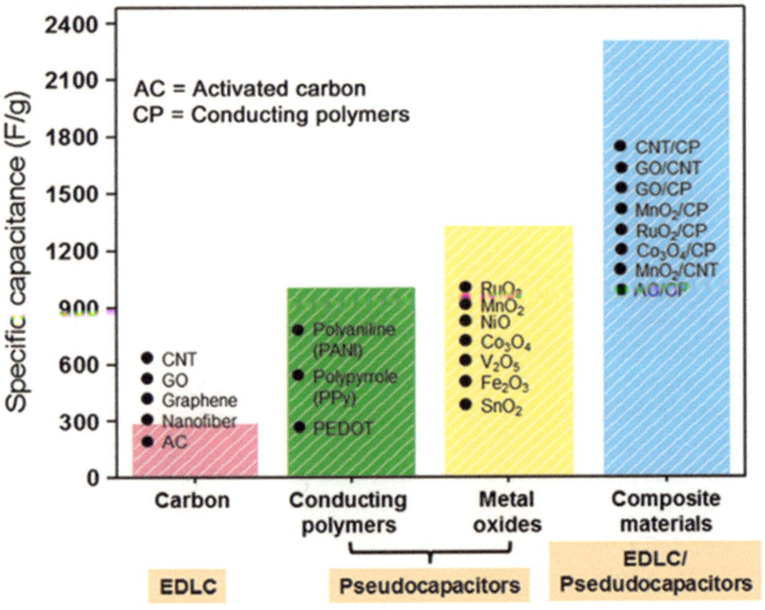

Figure 2.
Comparison of various materials according to their specific parameter for supercapacitor applications [21].

investigated potential range. The resulting response is similar to the electrochemical response of charging/discharging the double layer and therefore these reactions are called pseudocapacitive reactions. This property enables the application of these materials in supercapacitors. In order to further improve the properties of conducting polymers, attempts are being made to increase their electrical conductivity and porosity [21].

In **Figure 2** a comparative review of current supercapacitor electrode materials has been provided. Carbon materials have a specific capacitance of up to 300 F/g, while polymer and metal oxide materials can have over 1000 F/g. Composites of two or more of the above materials can have a very high specific capacitance of over 2000 F/g. Recently, the predominant approach is the development of binary or ternary nanocomposites of different capacitive materials to determine and optimize the structures and physical and mechanical properties of the electrode materials, in order to achieve improved supercapacitor performance. However, the properties of composite electrodes, in addition to the individual active components, also depend on the morphology and characteristics of the interphase [21].

Special, often complex technologies are developed for the production of superior electrode material. In **Figure 3** one of them is shown. A schematic diagram of the entire process of $MnNi_2O_4@MnNi_2S_4$ electrode materials is presented. Ni^{2+} and Mn^{2+} form Mn-Ni precursors in the reactor and are then calcined at high temperature to produce oxides. Then, under the influence of sodium sulphide, ion exchange is carried out at the appropriate temperature, i.e. oxygen is replaced by sulphur, which is less electronegative. A core-shell structure is formed without changing the morphology of the oxide. Metal ions react with KOH electrolyte in the Faraday redox reaction [22].

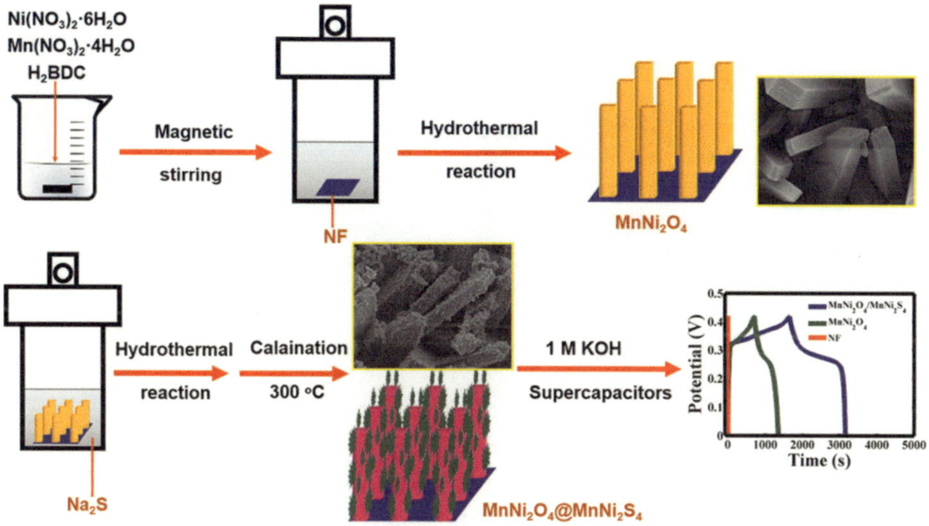

Figure 3.
The schematic diagram of the construction process for $MnNi_2O_4@MnNi_2S_4$ electrode [22].

4. Production, modeling, and characterization of supercapacitors

Supercapacitors fill a wide area between storage batteries and conventional capacitors. Both from the aspect of energy density and from the aspect of power density this area covers an area of several orders of magnitude. Materials, construction, production technology, and test methods are chosen depending on the part of the area covered by a certain type of supercapacitor.

4.1 Construction and production of SC

The construction of the laboratory prototype and zero series begins with the choice of material and type of supercapacitor (asymmetrical - polarized, or symmetrical - nonpolarized). The variant with a solid active material, which the authors of this chapter have often used, is shown in **Figure 4**. The connection of the active material

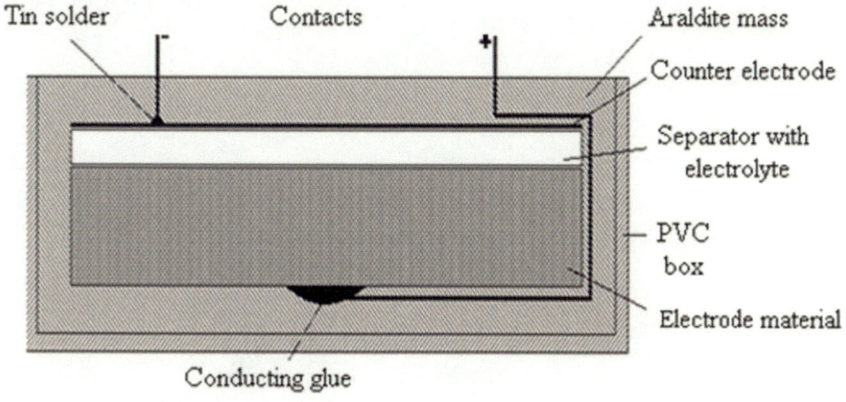

Figure 4.
Construction of supercapacitor prototype [23].

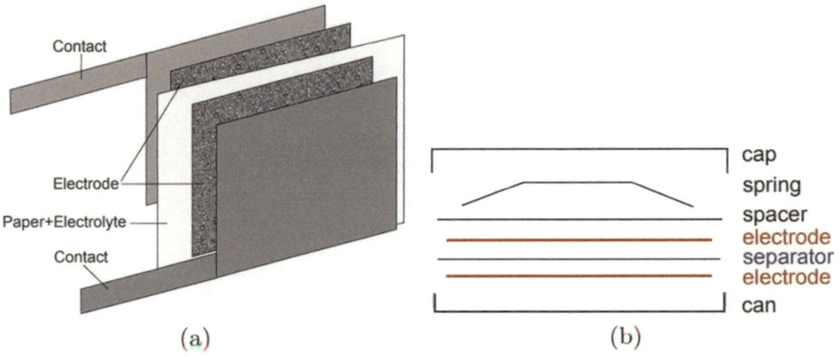

Figure 5.
Configuration of supercapacitors in various casings: (a) in laminated cells and glass cells; and (b) in coin cells [24].

with the power supply is achieved by using conductive silver glue. The role of a separator is provided by a Nafion foil soaked in a selected electrolyte.

Today, the active material is applied from a suspension, as well as the printing techniques that are also applied for micro SC (**Figures 5** and **6**).

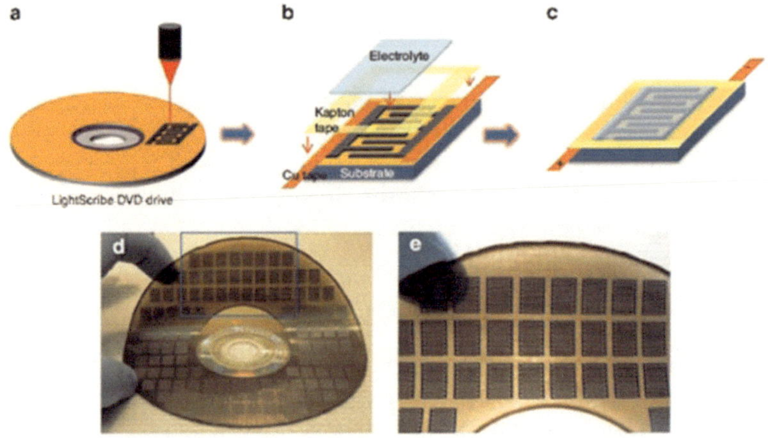

Figure 6.
(a–c) Schematic diagram of the fabrication process for micro-supercapacitors by laser scribing method. (d, e) Flexible micro-supercapacitors with high areal density [25].

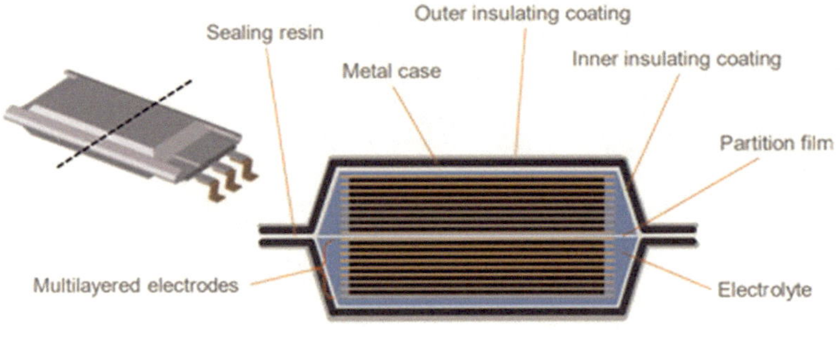

Figure 7.
Structure of Murata's supercapacitor (cross-section) [26].

Figure 8.
Supercapacitor engine start module [27].

After testing and optimization, a zero series is made, where the technological processes of production are elaborated (and often simplified) and only then the mass production could start. The appearance and structure of a commercial supercapacitor of low power and capacitance are shown in **Figure 7**, and for higher powers, voltages, and capacitances, it is shown in **Figure 8**.

Depending on the field of application and the set of parameters, the final structure, production technology, and housing are selected. If the supercapacitor is used as a replacement for a battery to power small consumers, in addition to the capacitance, it is very important that it has a low leakage current. In contrast, when powering larger consumers, it is much more important that the supercapacitor has a low series resistance due to losses at high currents. Thus, the type of supercapacitor is defined. As an example, in **Figure 9**. the types of supercapacitors are given for the different needs of power supply backup.

4.2 Modeling and characterization of SC

In order to predict the behavior in different conditions, electrochemical processes are modeled and simulated on the computer [2]. Most often, an equivalent electric circuit with two or more RC branches is taken as a model (**Figure 10**) [29].

For many years, the authors of this chapter have done research in this area and have developed their models. The simplified model will be presented here for the case of sulfide minerals in an aqueous solution of sulfuric acid. The equivalent circuit, shown in **Figure 11**, was adopted. Capacitors represent double-layer and diffusion capacitance, and resistors correspond to electrolyte resistance, diffusion resistance, and leakage current.

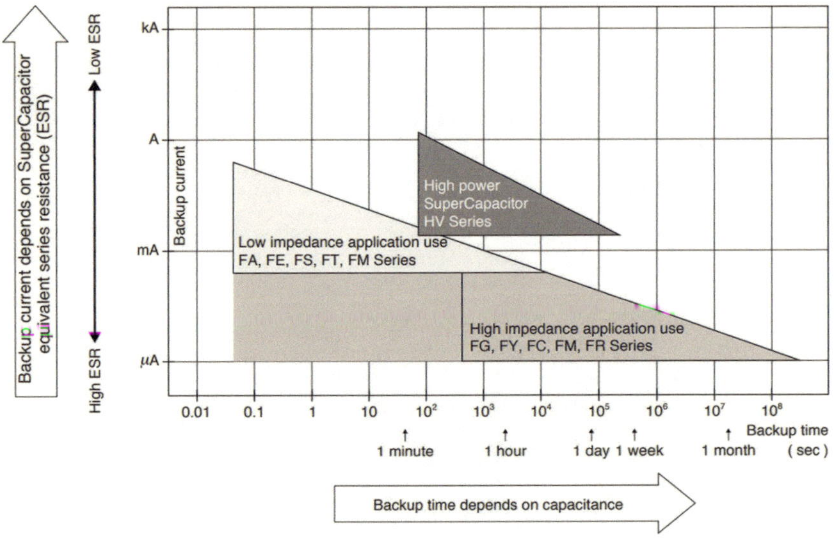

Figure 9.
Performance for supercapacitor selection [28].

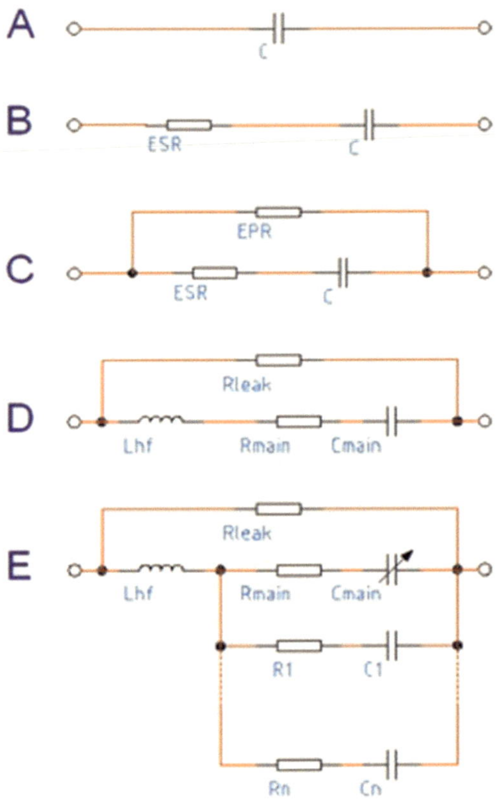

Figure 10.
EDLC models: A - ideal capacitor, B - series RC model, C - model B with added leakage resistor, D - model C with added high-frequency inductance component, and E - model D expanded with n-branch RC circuits and voltage-dependent main capacitance [29].

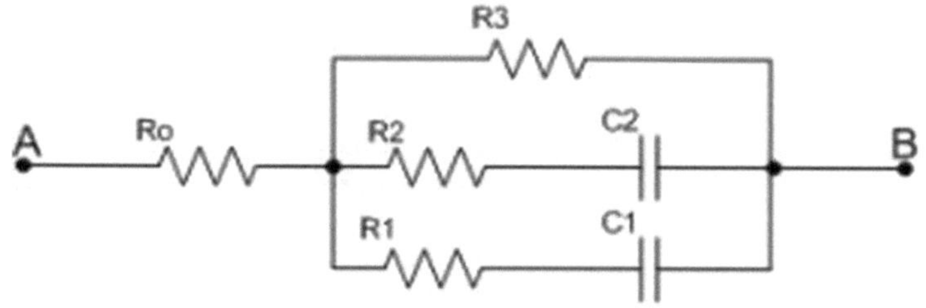

Figure 11.
The equivalent electric circuit.

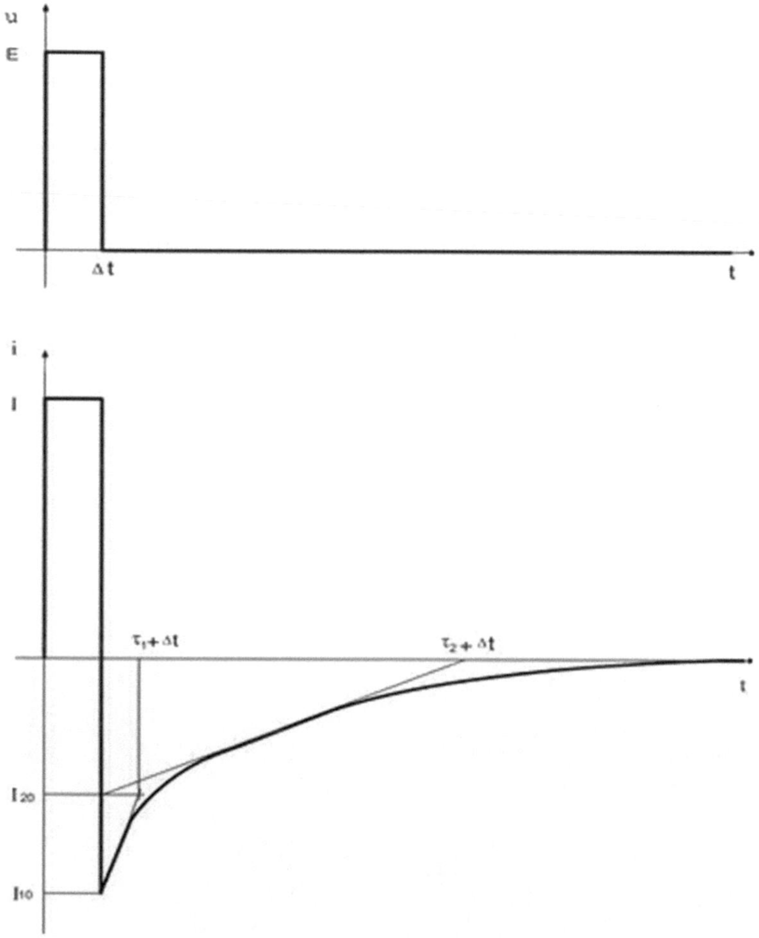

Figure 12.
The circuit excitation and the response.

For the assumed equivalent electric circuit of the observed electrochemical system and a short voltage pulse (of the order of 0.1 s for such systems), the response of the system (current in this case) will be as shown in **Figure 12**. The following parameters are:

$I = \frac{E}{R_0+R_{123}}$ - quasi-stationary charging current.
$R_{123} = \frac{R_1R_2R_3}{R_1R_2+R_1R_3+R_2R_3}$ - eq. resistance of parallel connection R1, R2, and R3.
$I_{10} = \frac{-U_{C_{10}}}{R_1+R_{023}}$ - initial discharge current.
$R_{023} = \frac{R_0R_2R_3}{R_0R_2+R_0R_3+R_2R_3}$ - eq. resistance of parallel connection R0 , R2, and R3.
$U_{C_{10}} = \frac{I_2 \bullet \Delta t}{C_1} \frac{R_2}{R_1+R_2}$ - initial discharge voltage of capacitor C1.
$I_{20} = \frac{-U_{C_{20}}}{R_2+R_{03}}$ - quasi-stationary discharge current.
$R_{03} = \frac{R_0R_3}{R_0+R_3}$ - eq. resistance of parallel connection R0 and R3.
$U_{C_{20}} = \frac{I_2\Delta t}{C_2} \frac{R_1}{R_1+R_2}$ - initial discharge voltage of capacitor C2.
$\tau_1 = (R_1 + R_{023})C_1$ - time constant of the first discharge phase.
$\tau_2 = (R_{21} + R_{03})C_2$ - time constant of the second discharge phase.

Based on the given analysis, the parameters of the electric circuit can be determined, and therefore, in the same manner, the physical parameters of the electrochemical system [30].

In order to check the provided model, an electrochemical system was formed. Several experiments were performed using this method and it has been verified and determined that it can be applied for rapid characterization of electrochemical systems. The experiments were carried out using a system for electrochemical tests based on a PC and the LabVIEW software package (**Figure 13**) [31]. The system is designed to cover most electrochemical tests in wide ranges, both at the level of the electrochemical cell, and at the level of the completed supercapacitor (**Figure 14**) [32].

Modeling can also be done on a physical level. For example, a finite element model for charge transport in conjugated polymers has been developed, but it is still being refined [33].

Figure 13.
System for electrochemical testing.

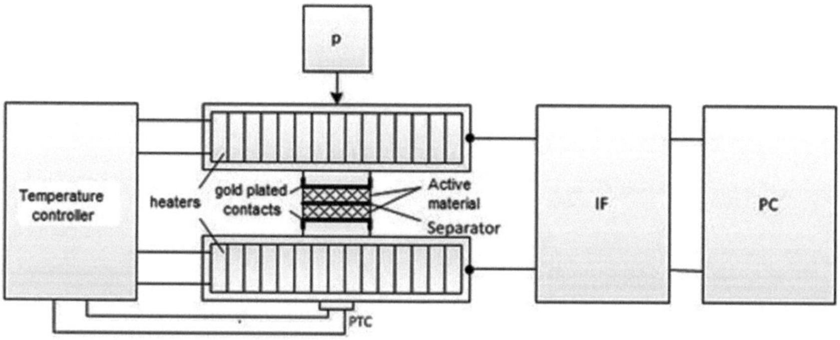

Figure 14.
Block diagram of the system for supercapacitor prototype characterization.

5. Use of supercapacitors

From the user's point of view, the most environmentally friendly form of energy is electricity. However, if we take into account the way in which this energy was obtained, then it fully retains its ecological advantage only if it originates from solar energy, wind, and wave energy, and to a considerable extent, from hydropower. In the latter case, namely, significant harmful ecological effects may occur due to disturbance of water regimes, such as surface waters. However, even when the electricity comes from the burning of fossil fuels or from nuclear reactions, problems related to the negative effects of by-products can be solved much more efficiently in one place than, for example, in all vehicles that move using the appropriate energy. One of the conditions for the use of electric energy in vehicles is the existence of such a device that would have high specific energy but at the same time a high specific power, which standard electric devices could not provide. The advent of supercapacitors has made this application much more realistic.

In the case of supplying consumers with energy that comes from solar or wind energy, it is necessary to have an appropriate method of energy storage for the period when there is no sun or wind. In this case, supercapacitors have an advantage over standard batteries because they can withstand a much greater number of charging and discharging cycles.

From a very large number of possible applications of supercapacitors, current examples will be listed where their characteristics are irreplaceable.

5.1 Energy harvesting

Monitoring of environmental parameters requires the installation of systems in inaccessible and dangerous terrains. After installation, the system is expected to have a lifetime as long as possible with minimal maintenance. In addition, energy consumption is directly related to the lifetime of a wireless sensor network (WSN). It is similar to other scattered and/or remote systems. Primary (non-rechargeable) batteries, despite the application of modern energy management algorithms, have the greatest impact on the limited lifetime of a wireless sensor node. Also, regular technical interventions in the field, primarily battery replacement, drastically increase the cost of maintenance. With the aim of increasing the life span and reducing maintenance costs, current research studies involve the use of secondary (rechargeable)

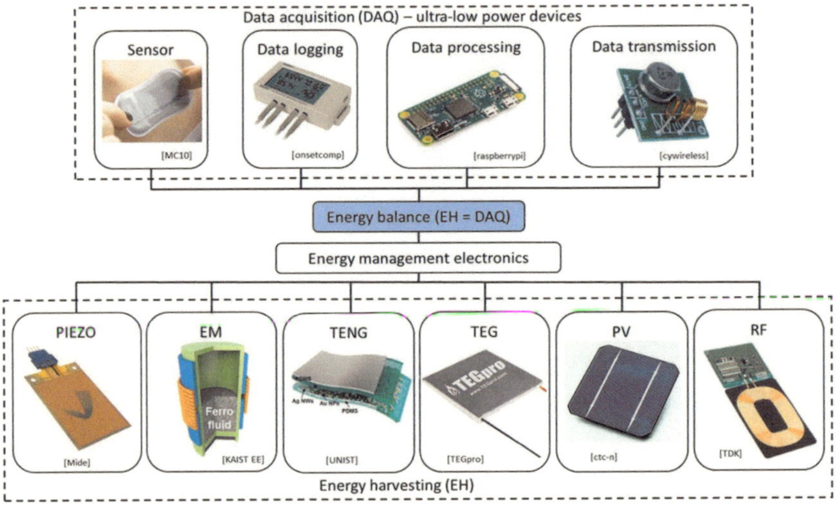

Figure 15.
Environmental energy sources [35].

batteries and the so-called collecting energy from the environment, that is, "energy harvesting" (EH), which contributes to WSM getting the self-powered prefix. Due to the characteristics of secondary batteries that degrade over time, the increase in lifetime is insufficient for multi-year monitoring of certain environmental parameters. This is the reason why, instead of rechargeable batteries, capacitors with very large capacities— supercapacitors—can be used to power the node. They represent reversible electrochemical systems, and they are increasingly used to power sensor nodes. For several reasons, supercapacitors are favorable for power supply, one of them being the exceptional scalability that allows increasing capacity and performance with increasing dimensions and weight. The characteristics of supercapacitors, such as high-power density, fast charging, large number of charging cycles, temperature stability, small equivalent series resistance, and low leakage current, favor the operation mode of most wireless sensor nodes. However, the lower energy density compared to batteries contributes to the fact that they are discharged relatively quickly and require frequent recharging. That is why it is necessary to provide a constant or at least intermittent source of energy in the natural environment. It can be a solar panel, piezo vibration transducer, thermoelectric generator, antenna, etc. [34]. Possible sources of environmental energy are shown in **Figure 15**.

5.1.1 Piezoelectric effect

Piezoelectric materials have the property of converting energy through the direct piezoelectric effect, the energy of mechanical deformations of the piezoelectric structure into the electric field, that is, voltage [36]. Two modes, that is, mode 33 and mode 31, are used in most developed EH applications (**Figure 16**). In both of these modes, the electric field, and thus, the generated voltage on the electrodes, are oriented along the direction of polarization 3, while external forces cause stresses in a single direction. In mode 33, this is the same direction as the stress (3), while in mode 31 the stresses are along the normal (1) [37].

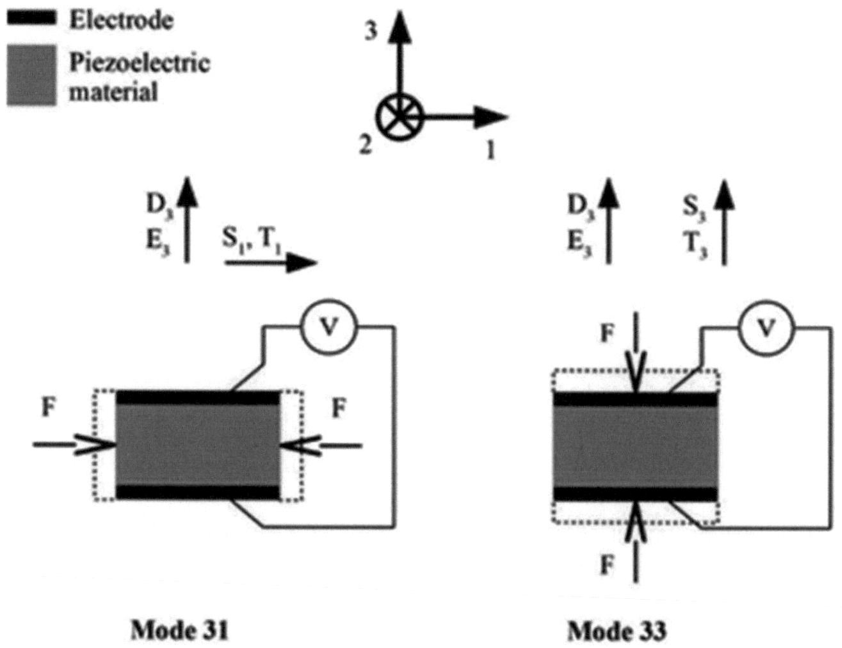

Figure 16.
Operational modes of piezoelectric material for EH applications [37].

5.1.2 Electromagnetic conversion

Electromagnetic induction, described by Faraday's law, is the creation of electromotive force (EMF), that is, voltage on an electric conductor in a changing magnetic field, a phenomenon that forms the basis of electric generators. The induced EMF is proportional to the strength of the magnetic field, the speed of the relative movement, and the number of turns of the conductor. If a conductor is connected to an electrical load, the current will flow, thus generating electricity. This system is often used as an effective tool for the realization of kinetic EH systems, where the relative displacement of the permanent magnet in relation to the coil is caused by the vibrations of the energy generation base [38].

5.1.3 Comparison of piezoelectric and EH devices with electromagnetic vibrations

The results of a thorough comparison of electromagnetic and piezoelectric vibration EH systems, with identical volumes, seismic masses, natural frequencies, quality factors, and excitation conditions, are given in [39]. Appropriate mathematical models of both types of vibration EH devices were used for the calculation of output voltages and powers (**Figure 17**) and depending on the intensity and frequency of harmonic dynamic excitation, the recommended configuration of the most efficient vibration EH systems [39].

5.1.4 Magneto-strictive effect

Physically in a way similar to the piezoelectric effect. A characteristic property of magneto-strictiveness is that the magneto-elastic coupling induces mechanical

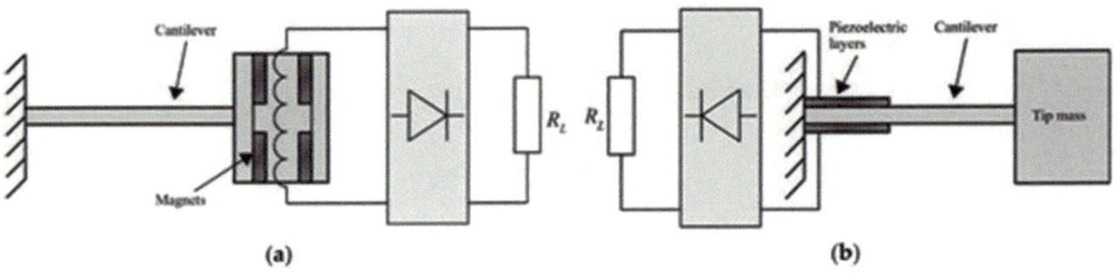

Figure 17.
Model of electromagnetic (a) and piezoelectric (b) vibration EH systems for output power analysis [39].

elongations if they are subjected to a magnetic field, while conversely, their magnetization will change due to changes in the applied mechanical stresses. This effect can be used in EH devices, however, an additional coil is required to obtain electrical energy [35].

5.1.5 Photovoltaic panels

Solar panels are easy to purchase and install. The price is getting lower, so they are a good choice for collecting energy from the environment. Since the panels provide a DC output, they can be plugged directly into the electronics' power system. The output power can be from mW to MW depending on the size of the panels as well as the intensity of the light they receive. However, the main limitation is that they need to be exposed to sunlight, so they can only work during the day. Batteries and/or supercapacitors are necessary for power supply at night. Energy storage is also necessary for cloudy or snowy days [40].

5.1.6 Thermoelectric systems for energy harvesting

In addition to mechanical energy, a temperature difference is also a very rich source of energy; therefore, often considered a viable option for the development of EH systems. To convert heat into electricity, thermoelectric generators (TEG) use the available temperature differences between two surfaces [35].

Thermoelectric devices for energy production or cooling consist of two types of thermoelectric materialsthermoelements. These are p-type and n-type semiconductors. Thermocouples are electrically connected in series and thermally connected in parallel. The working principle of TEG is shown in **Figure 18**.

5.1.7 Wind energy harvesting

Wind-based energy harvesting is increasingly pursued due to the ubiquitous nature of the source, as well as complementarity with other sources (e.g., solar). In particular, vortex-induced vibrations (VIV) are investigated [41].

Vortex-induced vibration energy harvesting mechanisms can be divided into five categories, namely, flutter, VIV, galloping, wake-galloping, and hybrid-type flow-induced vibrations. VIV energy harvesters are designed for regions with low wind speed [42].

5.1.8 Storaging of harvested energy by supercapacitors

Regardless of the source of clean renewable energy, it is necessary to have a circuit to store the energy generated from the energy harvesting source. When a DC voltage is

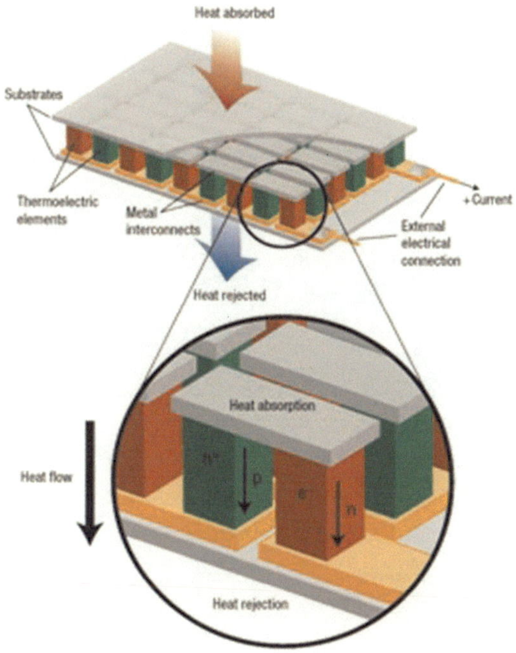

Figure 18.
Thermoelectric device principle [40].

applied to a discharged supercapacitor, it is charged, and thus stores electrical energy. Since these are small consumers (sensors, transmitters, and IoT), today's supercapacitor can often replace batteries and be more durable and environmentally friendly.

Most consumers require a higher operating voltage than a single supercapacitor can provide. In systems requiring higher voltages, supercapacitors are usually connected in series. In series-connected supercapacitors, however, a balancing circuit such as this, is required to distribute the voltages across the individual elements equally. There are two types of balancing circuits: passive balancing and active balancing [43].

A team of scientists from the American UCLA and the University of Connecticut [44] designed a system that is powered by electrical impulses from the human body. It is a "biological supercapacitor" that uses charged particles and ions, from the fluids in the human body. The device is not dangerous for the body and it can be used in pacemakers and other implants that require a power supply (**Figure 19**).

The block diagram of a system for collecting the energy of light radiation (natural or artificial) is shown in **Figure 20**. A supercapacitor with a capacity of 400 mF was used.

Another example of ultra-low power management, with a supercapacitor for energy storage (1.5 F) is shown in **Figure 21**. MOSFETs are used to rectify the output voltage of a wind energy harvester exposed to low wind. The proposed algorithm enables the monitoring of the maximum output power at time-varying wind speeds. A microcontroller was used to provide a source and sink impedance matching [46].

5.2 Smart cities

Following the smart city concept, supercapacitors have the potential to be involved in the creation of greener, sustainable, and efficient powering systems. One of the

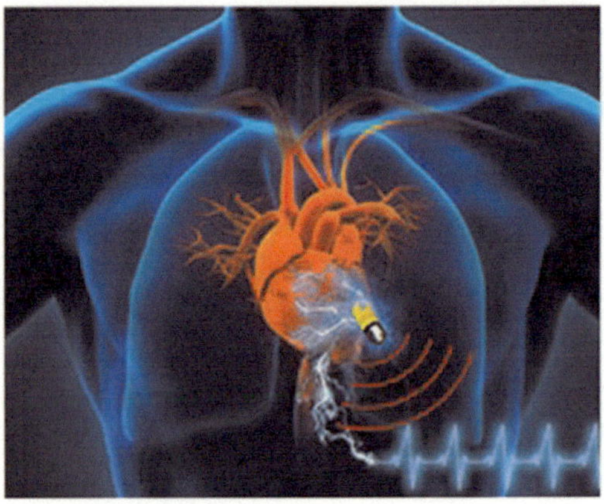

Figure 19.
Energy harvesting power management [44].

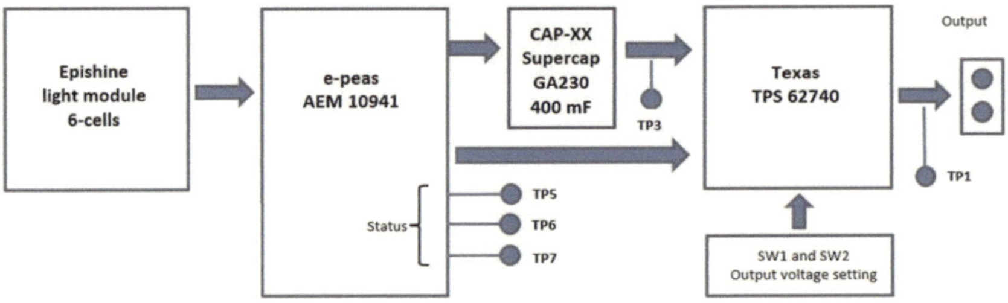

Figure 20.
Energy harvesting power management [45].

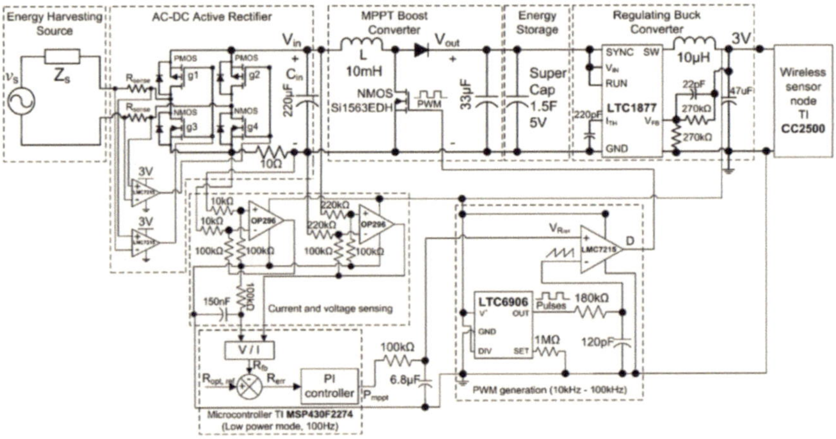

Figure 21.
Interface circuit for the harvester [46].

most prominent examples is public transport. By using the distributed energy sources in the urban smart environments, the power sources become DC based including the photovoltaics cell, and fuel cell, etc. As the urban environments are designed with many distributed power sources connected to the distribution lines, energy storage takes a significant place in the system. Battery energy storage systems and supercapacitor energy storage systems, as well as hybrid ones, may be installed both on large and small scales, which makes them the ideal fit for the smart city concept [47].

The smart city concept cannot be imaginable without sensor networks and Internet of Things devices and applications. As the energy requirement in sensor devices is increasing, the energy has to be stored for the blackout periods. Considering that the batteries are not a permanent solution, the supercapacitors serve as a solution for high-energy storage applications that require high-voltage and high-current drive [48]. Recent studies show that the supercapacitors are well suited for a wide range of applications, such as IoT, consumer products, white goods, office automation, long-term battery backup, and energy harvesting [48]. In order to overcome the powering issues that may occur at the remote nodes, as well as in the extreme weather conditions, fully functional IoT devices have been designed based on energy harvesting with supercapacitors and batteries as storage elements [47].

5.3 Smart grid

In recent years, the economic trends have been dictating the renewable energy sources generation of electric power. Therefore, the concept of the microgrid has been introduced as an off-grid or grid-connected energy system that can work independently or collaboratively with other microgrids [49]. In general, such a system can provide electric power either from a single source or multiple sources, such as wind and solar energy, adding energy storage to the system [50, 51].

The supercapacitors are being used to regulate the microgrid voltage and to improve the system stability. In recent studies, the supercapacitor provides the error component of the battery current in the proposed control scheme. This is an addition to the microgrid, as it improves the microgrid voltage regulation capability, as well as extends the battery lifetime [52, 53].

5.4 Energy systems

Supercapacitors are increasingly used in both AC power systems (EES) and DC power sources. With the development of the voltage balancing technique of serially connected SCs, a great improvement in the HVDC transmission system is expected. The large capacity of SC provides enough energy storage for small consumers in a short time, and their main advantage in energy systems is high power density, so they can cover large consumption peaks. In combination with power electronics circuits, SCs can inject energy into the EES at the right moment, thus opening a whole new field of development of circuits and control algorithms. A large field of application of SC in DC power supplies is low-pass filters with previously unimaginable parameters.

5.4.1 Quality of electricity in AC systems

Experts realized a long time ago that the quality of electricity affects the quality of work and life. Many norms from that area were applied, but the real progress started at the end of the last century. The problem began to be approached globally. CENELEC (European Committee for European Standardization) was established [54]. Most European countries have accepted the CENELEC standard EN 50160 [55] for voltage monitoring at the point of delivery to the consumers under normal conditions.

The LEM corporation (NORMA, ELMES, ELSIS, and HEME) developed a series of measuring devices MEMOBOX for monitoring the quality of electricity, and thus, began a new practice supported by the EN 50160 standard, but also by newly created standards (e.g., IEC 1000-3-6/71).

As connecting the national and European power systems is a necessity, it is also necessary to adapt local legislation and standardization in this area. The specificity of electricity is that its quality is influenced to a greater extent by consumers (non-linear loads) than by producers. That is why the consumer is, to a considerable extent, a partner of the supplier in ensuring the quality of electricity. At the same time as electrical energy becomes dirty, the consumer is also sensitive to this kind of contamination. From the point of view of the application of supercapacitors for the elimination of short-term disturbances, the following terms should be highlighted [55]:

5.4.2 Frequency change

Under normal operating conditions in the distribution network connected to the power system, the ten-second mean value of the frequency during 99.5% of each week must be within 50 Hz ± 1% and 50 Hz + 4% / - 6% in the remaining 0.5% of the week.

In disconnected (island) networks, the limits are 50 Hz ± 2% during 95% of the week and 50 Hz ± 15% during the remaining 5% of the week.

5.4.3 Flicker

The need to define and measure that parameter is created from the fact that the change in light intensity in the working or living environment negatively affects people's health, that is, their work and other efficiencies. Headaches, nervousness, depression, and vision impairment, etc. occur. Flicker is defined as follows: If there are 100 people in a room under equal conditions, and if the light intensity changes so much that 50 of the 100 people notice it, the flicker is said to have an intensity of 1. Flicker is a consequence of amplitude modulation supply voltage with frequencies in the range from 1 to 33 Hz, where the amplitude is a direct function of that frequency. For example, at a frequency of 8 Hz, the nominal voltage fluctuation amplitude is about 0.256 % of the nominal value (e.g., 0.59 V from 230 V~).

5.4.4 Voltage failure

Voltage failures occur most often due to faults in the consumer's facilities or in the public distribution network. They are defined as follows: failure (partial loss of voltage) is sudden (unpredicted), short-lived (from 10 ms, up to 1 minute) reducing the supply voltage to one of the values in the range of 90%, and up to 1% of the nominal voltage, after which the nominal voltage is restored. The permissible guideline number of voltage drops during one year is ranged from 10 to 1000. Most of them must have a duration of less than 1s and an amplitude of less than 60% of the nominal voltage.

5.4.5 Power failure

It is a state in which the voltage at the point of transmission is less than 1% of the nominal voltage. The following power interruptions are distinguished:

- Planned interruptions of supply, about which consumers are informed in advance, in order to enable the planned works to be carried out in the network, and

- Accidental interruptions caused by permanent or transient disturbances they usually appear in conjunction with other disorders.

The following random interruptions are distinguished:

- Long-term interruptions (longer than 3 minutes), caused by permanent failure

- Short-term supply interruptions (up to and including 3 minutes), caused by a transient fault.

Figure 22 shows an example of a short-term power failure.

Figure 23 shows the configuration of the system supercapacitor in the control area of the power system. The converter (VSC) consists of a rectifier/inverter with 6-pulse control and pulse width modulation (PWM) with an IGBT bridge. The PWM converter and the DC-DC converter (chopper) are connected by a DC link capacitor. A bidirectional DC-DC converter operates in step-up mode if electrical power is supplied to the supercapacitor bank from the power system. Smoothing inductance is used for current transfer and filtering [56].

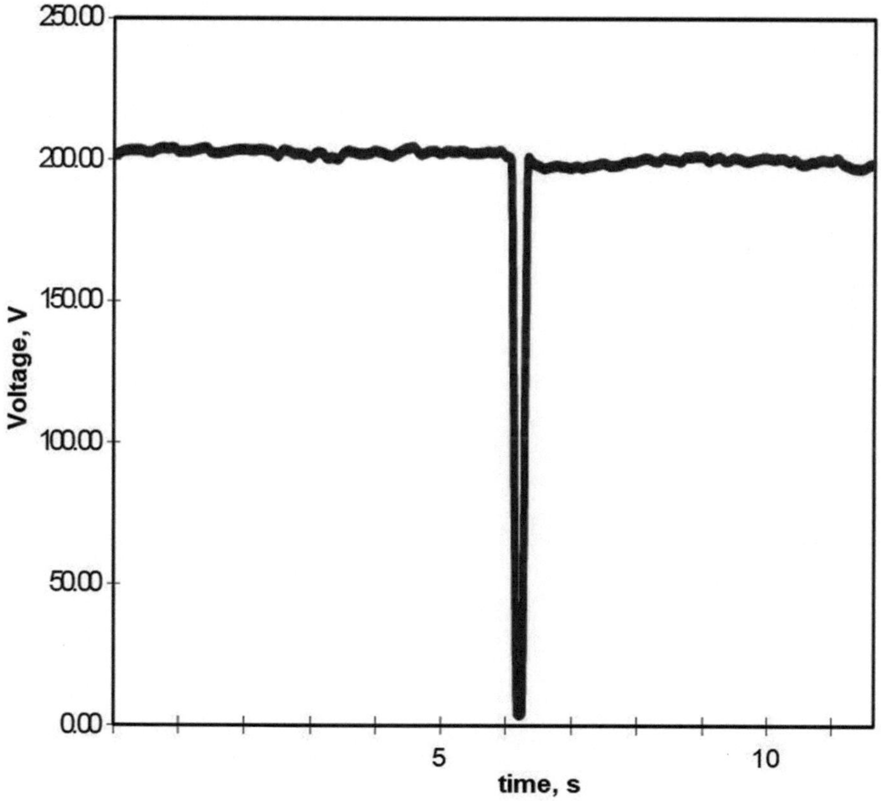

Figure 22.
Example of a short-term power failure [54].

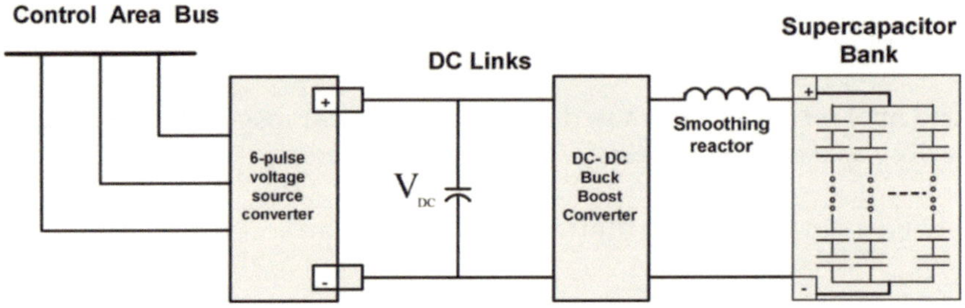

Figure 23.
Configuration of supercapacitor bank in the control area [56].

5.4.6 Supercapacitors in DC systems

Supercapacitors are most often polarized, and due to their enormous capacitance, they are most often used in DC systems. The most application is in the accumulation of electrical energy to cover shorter production stops and large consumption peaks, but they are widely used in DC voltage filtration and other purposes in power electronics circuits.

An example of an independent photovoltaic system with supercapacitors for energy storage is shown in **Figure 24**.

The operating voltage of the supercapacitor can be kept within range by properly sizing the supercapacitor and monitoring the upper and lower limits of the charge and discharge controller (SOC). The rate of change in the supercapacitor is proportional to the charge current isc. To extract the maximum available power from a PV panel, it is necessary to operate the PV at its maximum power point (MPP). The MPP monitoring

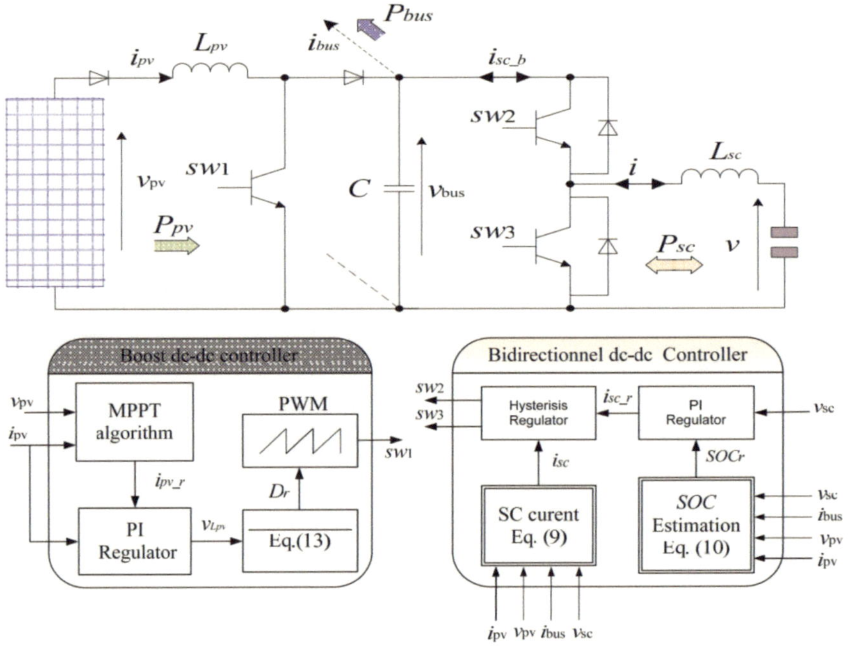

Figure 24.
PV panel and supercapacitor controller system scheme [57].

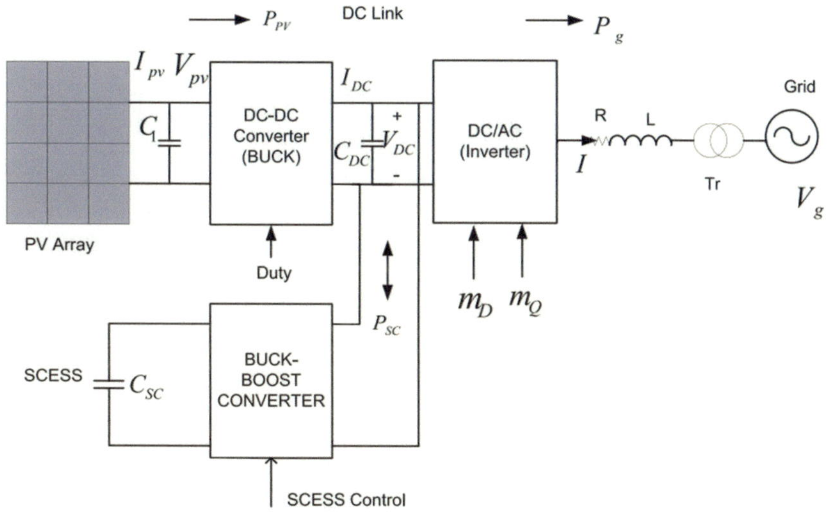

Figure 25.
PV panel and supercapacitor connected on grid [58].

device is a high-frequency DC-DC converter, a chopper-voltage booster inserted between the PV panel and the DC bus [57].

An example of a PV system connected to the distribution network, where supercapacitors are also used, is shown in **Figure 25**. The power generated by the PV panel is connected to the grid using buck converters and inverters. A buck chopper is used to regulate the variable voltage from the PV panel. Its duty cycle is continuously adjusted to instantly locate the maximum power output from the PV panel at varying irradiance and temperature. To minimize the fluctuation in the generated power, a bidirectional buck chopper is used to connect the energy storage to the DC link. The energy storage (supercapacitor bank) is continuously charged and discharged by a buck chopper to absorb or release the required power between generated and transmitted to the grid. The step-up chopper controls the supercapacitor voltage and the DC link voltage. An independent VSC active and reactive power regulator was implemented to inject available power into the AC network [58].

A hybrid system for electrical storage based on supercapacitors and batteries is shown in **Figure 26**. In a hybrid system, the peak load power is supplied from the supercapacitor and the batteries provide lower constant power for a longer time.

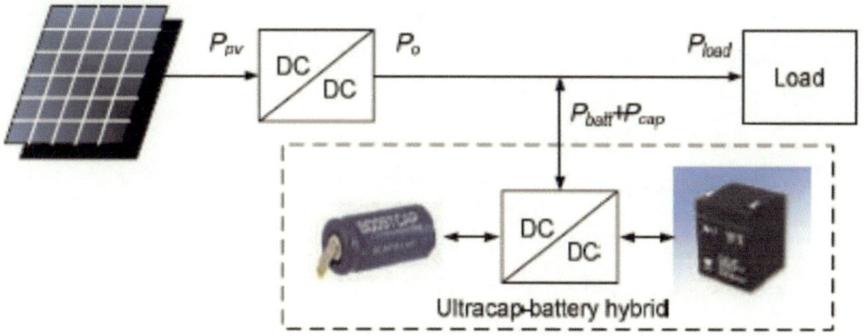

Figure 26.
Supercapacitor-battery hybrid energy storage in PV system [59].

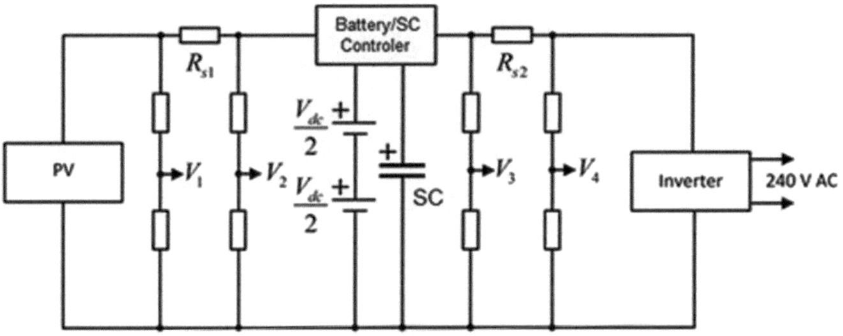

Figure 27.
Schematic of PV/batt/SC system.

The authors of this chapter have designed a sample PV system with supercapacitors and batteries for energy storage (**Figure 27**). A system for monitoring energy parameters was developed, and several algorithms of energy management and MPPT were also implemented. In **Figure 28** data acquisition block diagram is given. Analog channel AI0 monitors the temperature of the PV module in which the LM35 chip is installed. Channels AI1 to AI4 monitor voltages V1 to V4 (from **Figure 27**). **Figure 29** shows the basic LabVIEW monitoring application. Based on the measured voltages, the currents and other relevant parameters are calculated. Additionally, all the parameters and data are displayed, saved, and taken to further processing (implementation of the given algorithms).

The authors of this chapter have also set up a system with supercapacitors for injecting energy into the DC link of the self-excited asynchronous generator - rectifier - DC link with the supercapacitor - inverter - asynchronous motor system. Engine start-up and optimal energy management were tested. Due to the relatively high operating voltage (up to 300V DC), a bank of series-connected supercapacitors with passive voltage balancing was used, the efficiency of which was tested by thermal imaging (**Figure 30**). It is obvious that at point Sp1 the heating is significantly increased (58.40C), which means that passive voltage balancing should be improved (reduce the resistance of parallel resistors).

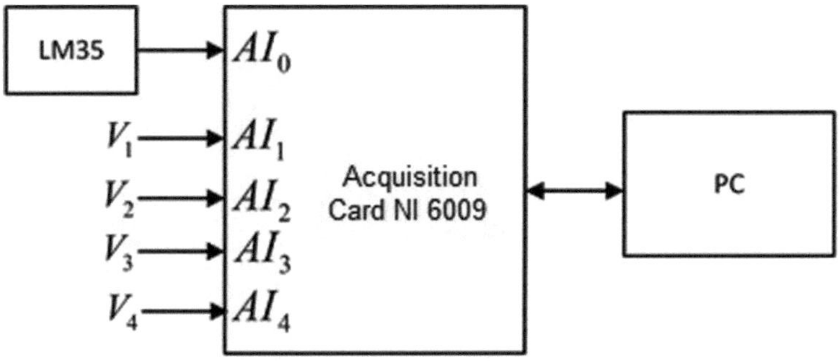

Figure 28.
Acquisition block diagram.

DOI: http://dx.doi.org/10.5772/intechopen.106705

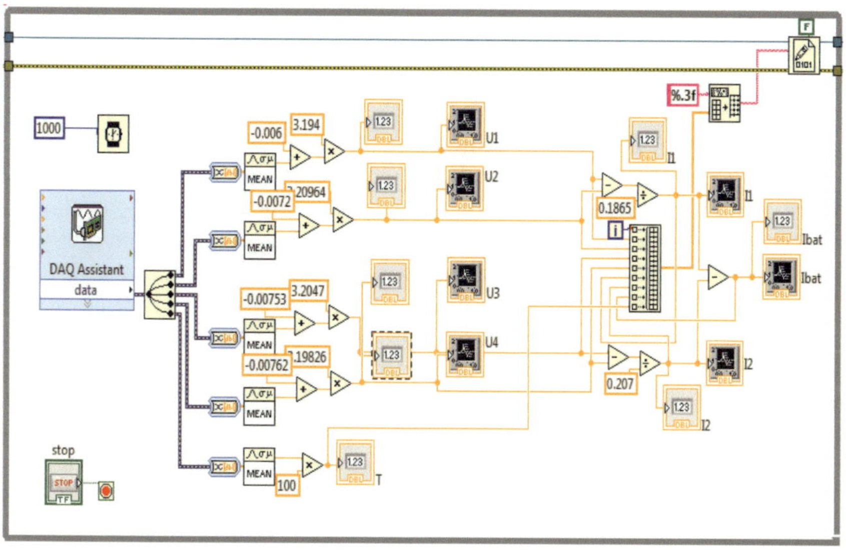

Figure 29.
LabVIEW block diagram of the monitoring system.

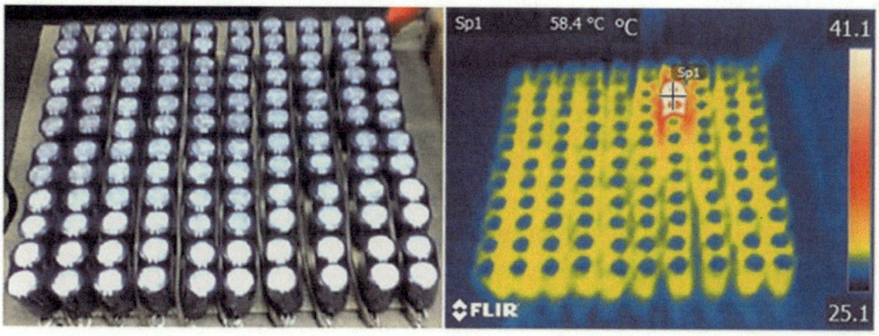

Figure 30.
Photo and thermal image of serially connected supercapacitors.

5.5 Electric vehicles

Vehicles with electric drive represent one of the most significant ecological advances, bearing in mind the prevalence of this type of contamination of nature. In the world, there is also interest in hybrid vehicles that have lower fuel consumption and significantly lower emission of harmful products compared to classic vehicles. In the most general form, hybrid vehicles can be described as vehicles that use a combination of technologies for energy production and storage. It combines the good features of conventional vehicles (long range and acceleration and excellent fuel supply network) and electric vehicles (zero emissions, quiet operation, and use of braking energy).

Time has shown that it is not necessary to immediately build a complete network of charging stations in order to increase the sale of electric vehicles since users are ready to charge the batteries of their vehicles at home. The possibility of supermarkets, parking garages, and restaurants offering charging stations to customers is mentioned as the next step.

5.5.1 Application of supercapacitors in EV

Certain characteristics of supercapacitors make these devices suitable for purposes in which a combination of high specific energy and high specific power is required, or a long service life expressed by the number of charging and discharging cycles. Namely, supercapacitors retain the positive property of standard capacitors that they can achieve an almost unlimited number of charging and discharging cycles.

From the point of view of the application, there are several groups of supercapacitors. Depending on the place of application, different characteristics of the supercapacitor come to the advantage more or less. Some of them are of crucial importance for the selection of capacitors, and some may be unimportant. The strictest requirements are set for capacitors used in electric traction, that is, in electric vehicles. Batteries with a capacity of several hundreds of farads and with an operating voltage of several hundred volts are already being made. In addition to high capacitance and relatively high operating voltage, these capacitors must have high specific energy and power (due to limited space in the vehicle). In terms of specific power, they have a great advantage over storage batteries, but they are, therefore, incomparably weaker in terms of specific energy. That is why the ideal combination is a combination of accumulator and capacitor batteries (**Figure 31**). In steady mode (normal traction), the vehicle's engine is powered from the battery, and during sudden acceleration from the supercapacitor. Especially important is the fact that during sudden braking, all mechanical energy can be returned to the system by converting it into electricity only with the presence of a supercapacitor with high specific power. For the above reasons, the internal resistance of such supercapacitors must be extremely low. The leakage current is not relevant [60, 61].

The wheels are driven by a regulated electric drive. Regulated electric motor drives are developing very quickly and are placing ever stricter demands on speed (and position) and torque regulation before designers. From the energy point of view, it is desirable that their participation be as large as possible, because, by setting the speed of the drive to the optimal or necessary one, it is possible to save on the energy consumed.

A typical variable speed electric motor drive contains [62]: A big challenge with electric vehicles is the necessary increased operating voltage to power the electric motor (due to the reduction of losses), so supercapacitors are connected in large series strings. This reduces the equivalent capacitance, and also leads to voltage balancing

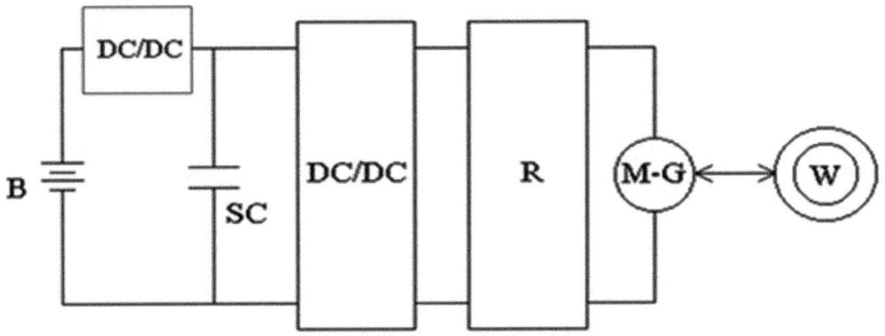

Figure 31.
Block diagram of an electric vehicle with a hybrid power supply.B - accumulator, SC - supercapacitor; DC/DC - converters of direct voltage; R - regulator; M-G - motor-generator (depending on the operation mode); and W - wheels.

problems on individual cells. The first problem is solved by connecting several series strings in parallel, and the second by passive (cheap and bad) or active (expensive and excellent) voltage balancing.

- Between the source and the motor, a converter that adapts the characteristic sizes of the source: frequency, voltage, current, and number of phases to the needs of the motor. By dosing that energy, engine control is also achieved;

- Between the engine and the load, a mechanical transmission that adjusts the speed and torque of the engine to the speed and torque of the working mechanism (load);

- Information from all the mentioned elements (source, converter, motor, transmission, and from the load) is collected by the regulator (controller), which is based on the set (desired) parameters, performs automatic control of the drive

In an asynchronous motor, at a constant frequency and amplitude of the supply voltage, the rotor speed depends on the load moment, which requires complicated control algorithms in cases where precise speed and/or position control is required. This phenomenon is a consequence of the principle of operation of an asynchronous motor, which is electromagnetic induction and requires a difference in speed between the rotor and the rotating magnetic field generated by the stator in order for the electromagnetic torque to exist. The electronics that realize the aforementioned algorithms were expensive, therefore, it made it difficult to use asynchronous motors for such purposes. However, today, with the low cost of electronic components and the use of computers in the realization of regulation algorithms, they are increasingly used.

The introduction of supercapacitors and power electronics assemblies based on DC voltage interfaces lead to a significant improvement in the performance of electric vehicles, such as acceleration, use of braking energy, and reduction of dimensions. Blocks of high-power supercapacitors are now also installed in large vehicles (buses and rail vehicles). In **Figure 32** a complete green energy system is presented, where supercapacitors play an important role.

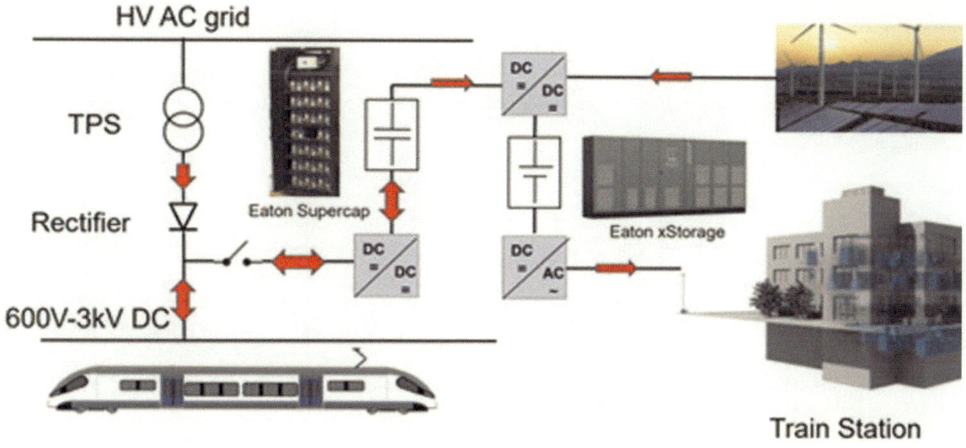

Figure 32.
Green transport [63].

5.6 Power electronics

Supercapacitors are, and in the future will be, increasingly used in power electronics assemblies of medium power, where they serve as reservoirs of electrical energy in the transition mode. There is a real possibility that they will soon replace bulky inductances, which are also huge sources of electromagnetic interference. In such an application, the supercapacitor must have both high capacitance and a relatively high operating voltage (which implies regular binding of the cells and all related problems). The internal resistance must be quite small, and the leakage current is not of major importance.

In **Figure 33** a schematic of a buck-boost converter with a supercapacitor for accepting braking energy is given.

In **Figure 34** the scheme of the SMPS converter with supercapacitor is given and **Figure 35** shows the supercapacitor in the inverter circuit.

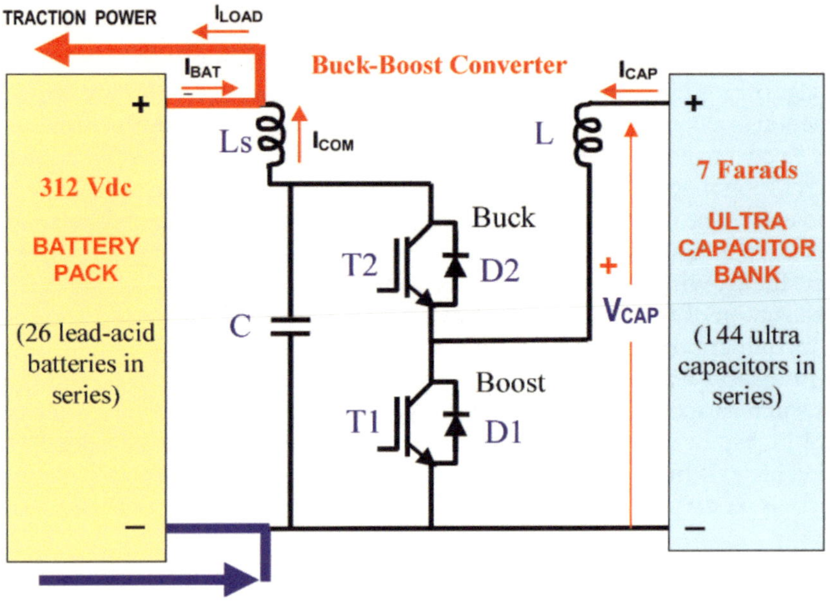

Figure 33.
A typical modern supercapacitor system with a bi-directional converter with SC [64].

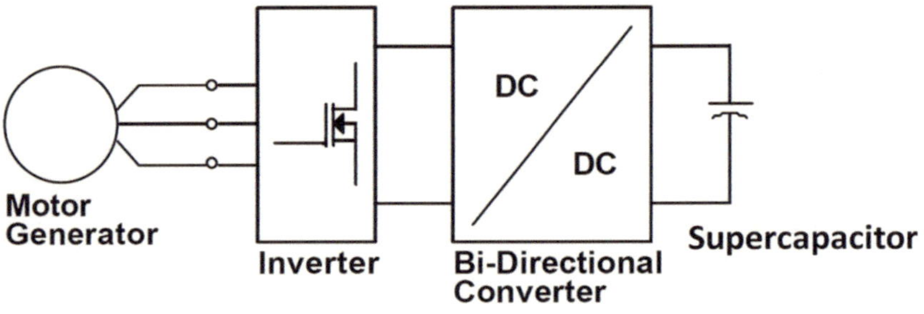

Figure 34.
Supercapacitor system with SMPS converter [65].

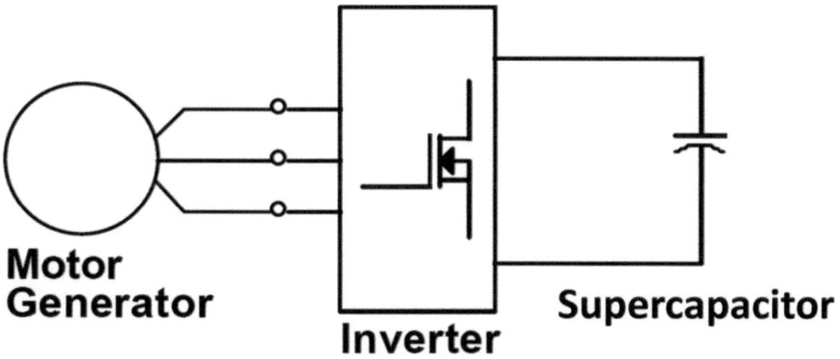

Figure 35.
Supercapacitor system with switch mode rectifier in the inverter [66].

6. Economy efficiency

Nowadays, the industry is focusing on improving product performance and reducing production costs. The supercapacitor components improve themselves, not only in the manufacturing process and technology, but in the direction to find stable and effective electrode and electrolyte materials, and to improve the performances, as well as to reduce the cost [8].

Following their properties, various industries and applications use supercapacitors, since they bring features, such as a safe, eco-friendly, and economical source of energy to the industry. It has been noted that the global supercapacitor market is supposed to attain US$ 8.3 billion by 2025. This market is predicted to enhance at a compound annual growth rate (CAGR) of 30% until 2025 [65, 66]. As provided in the recent reports, the cost of the material is a major constraint of market growth. Even though these disadvantages inhibit the supercapacitors' market growth, there is a high number of research groups that are focusing on minimizing the overall cost, and improving the supercapacitors adoption rate in the current market [66, 67]. CO_2 emission regulations that have been regulated in many countries worldwide, also have stimulated the usage of supercapacitors in the industry. This green agenda secures the future development of the supercapacitors, improvement of their properties, and their further research [13].

Acknowledgements

This research has resulted from the Projects funded by the Ministry of Education, Science and Technological Development, Government of the Republic of Serbia, grant numbers 451-03-9/2021-14/ 200135, 451-03-9/2022-14/ 200131, and 451-03-9/ 2022-14/ 200223.

Author details

Zoran Stevic[1*] and Ilija Radovanovic[2]

1 Technical Faculty in Bor, School of Electrical Engineering, D University of Belgrade, Bor, Serbia

2 Innovation Center of School of Electrical Engineering in Belgrade, School of Electrical Engineering, University of Belgrade, Belgrade, Serbia

*Address all correspondence to: zstevic@live.com

IntechOpen

© 2022 The Author(s). Licensee IntechOpen. This chapter is distributed under the terms of the Creative Commons Attribution License (http://creativecommons.org/licenses/by/3.0), which permits unrestricted use, distribution, and reproduction in any medium, provided the original work is properly cited.

References

[1] Conway BE. Electrochemical Supercapacitors. New York: Kluwer Academic/Plenum Publishers; 1999

[2] Stevic Z. [Ph.D thesis] University of Belgrade, School of Electrical Engineering, Belgrade. 2004

[3] Xiong S et al. A high-performance hybrid supercapacitor with NiO derived NiO@ Ni-MOF composite electrodes. Electrochimica Acta. 2020;**340**:135956

[4] Adib K et al. Sonochemical synthesis of Ag2WO4/RGO-based nanocomposite as a potential material for supercapacitors electrodes. Ceramics International. 2021;**47**(10):14075-14086

[5] Park J et al. Graphene-based two-dimensional mesoporous materials: Synthesis and electrochemical energy storage applications. Materials. 2021; **14**(10):2597

[6] Tie D, Huang S, Wang J, Zhao Y, Ma J, Zhang J. Hybrid energy storage devices: Advanced electrode materials and matching principles. Energy Storage Materials. 2018;**21**:22-40

[7] Zuo W, Li R, Zhou C, Li Y, Xia J, Liu J. Battery-Supercapacitor Hybrid Devices: Recent Progress and Future Prospects. Advanced Science. 2017;**4**(7): 1600539

[8] Shifei H, Zhu X, Sarkar S, Zhao Y. Challenges and opportunities for supercapacitors. APL Materials. 2019;**7**: 100901

[9] Xu Q, Wei C, Fan L, Rao W, Xu W, Liang H, et al. Applied Surface Science. 2018;**460**:84-91

[10] He Y, Chen W, Li X, Zhang Z, Fu J, Zhao C, et al. Freestanding Three-Dimensional Graphene/MnO2 Composite Networks As Ultralight and Flexible Supercapacitor Electrodes. ACS Nano. 2012;**7**(1):174-182

[11] Afif A, Rahman SMH, Azad AT, Zaini J, Islan MA, Azad AK. Advanced materials and technologies for hybrid supercapacitors for energy storage – A review. Journal of Energy Storage. 2019; **25**:100852

[12] Brandon NP et al. UK Research Needs in Grid Scale Energy Storage Technologies. London, U.K: Energy Storage Res. Netw. Eng. Phys. Sci. Res; 2016

[13] Berrueta A, Ursúa A, Martín IS, Eftekhari A, Sanchis P. Supercapacitors: Electrical characteristics, modeling, applications, and future trends. IEEE Access. 2019;7:50869-50896. DOI: 10.1109/ACCESS.2019.2908558

[14] Misra A. Energy storage for electrified aircraft: The need for better batteries, fuel cells, and supercapacitors. IEEE Electrification Magazine. 2018; **6**(3):54-61. DOI: 10.1109/ MELE.2018.2849922

[15] Banavath R, Nemala SS, Kim S-H, Bohm S, Ansari MZ, Mohapatra D, et al. Industrially scalable exfoliated graphene nanoplatelets by high-pressure airless spray technique for high-performance supercapacitors. FlatChem. 2022;**33**: 100373

[16] Lakra R, Kumar R, Sahoo PK, Thatoi D, Soam A. A mini-review: Graphene based composites for supercapacitor application. Inorganic Chemistry Communications. 2021;**133**:108929

[17] Olabi AG, Abdelkareem MA, Wilberforce T, Sayed ET. Application of

graphene in energy storage device–A review. Renewable and Sustainable Energy Reviews. 2021;**135**:110026

[18] Stević Z, Rajčić-Vujasinović M. Chalcocite as a potential material for supercapacitors. Journal of Power Sources. 2006;**160**:1511-1517

[19] Rajčić-Vujasinović M, Stević Z, Bugarinović S. Electrochemical characteristics of natural mineral covellite. Open Journal of Metal. 2012; **2**(3):60-67

[20] Yang C, Liang J, Xue L, Yue L, Liu Q, Lu S, et al. Recent advances in perovskite oxides as electrode materials for supercapacitors. Chemical Communication. 2021;**57**:2343

[21] Shown I, Ganguly A, Chen L-C, Chen K-H. Conducting polymer-based flexible supercapacitor. Energy Science & Engineering. 2015;**3**(1):2-26

[22] Lv X, Feng L, Lin X, Ni Y. A novel 3D MnNi2O4@MnNi2S4 core-shell nano array for ultra-high capacity electrode material for supercapacitors. Journal of Energy Storage. 2022;**47**:103579

[23] Stević Z, Rajčić-Vujasinović M, Bugarinović S, Dekanski A. Construction and characterisation of double layer capacitors. Acta Physica Polonica A. 2010;**117**:228

[24] Andres B. Paper-based Supercapacitors. Sweden: Mid Sweden University; 2014

[25] El-Kady MF, Kaner RB. Scalable fabrication of high-power graphene micro-supercapacitors for flexible and on-chip energy storage. Nature Communication. 2013;**4**:1475

[26] https://www.murata.com/~/media/webrenewal/products/capacitor/edlc/

[27] https://www.skeletontech.com/skelstart-ultracapacitor-engine-start-module

[28] https://www.tokin.com/english/product/pdf_dl/supercapacitors.pdf

[29] Čihak T, Jakopović Ž. Supercapacitors in power converter DC link. In: 2011 Proceedings of the 34th International Convention MIPRO. 2011

[30] Stevic Z, Rajcic-Vujasinovic M, Radovanovic I, Nikolic V. Modeling and sensing of electrochemical processes upon dirac potentiostatic excitation of capacitive charging/discharging. International Journal of Electrochemical Science. 2015;**10**:228

[31] Stevic Z, Stevic M, Radovanovic I, Stolic P, Milesevic M, Marjanovic M, et al. Computer-controlled voltage/current source and response monitoring system for electrochemical investigations. International Journal of Electrochemical Science. 2021;**16**. Article ID: 210659 1-14

[32] Stević Z, Rajčić-Vujasinović M, Panić V, Nikolić V. Uređaj za ispitivanje prototipova superkondenzatora. INFOTEH-Jahorina. 2015;**14**:61

[33] Wang X, Shapiro B, Smela E. Development of a model for charge transport in conjugated polymer. Journal of Physical Chemistry C. 2009;**113**: 382-401

[34] Mihajlović Ž. Wireless Sensor Network Node with Energy Harvesting for Monitoring of Environmental Parameters. Novi Sad: University of Novi Sad; 2018

[35] Zelenika S et al. Energy harvesting technologies for structural health monitoring of airplane components—A review. Sensors. 2020;**20**:6685

[36] Bowen CR, Kim HA, Weaver PM, Dunn S. Piezoelectric and ferroelectric materials and structures for energy harvesting applications. Energy Environmental Science. 2014;7:25-44

[37] Ambrosio R, Jimenez A, Mireles J, Moreno M, Monfil K, Heredia H. Study of piezoelectric energy harvesting system based on PZT. Integrated Ferroelectric. 2011;**126**:77-86

[38] James EP, Tudor MJ, Beeby SP, Harris NR, Glynne-Jones P, Ross JN, et al. An investigation of self-powered systems for condition monitoring applications. Sensor Actuators A-Physics. 2004;**110**:171-176

[39] Hadas Z, Smilek J, Rubes O. Analyses of electromagnetic and piezoelectric systems for efficient vibration energy harvesting. In: Fonseca L, Prunnila M, Peiner E, editors. Smart Sensors, Actuators, and MEMS VIII, Proc SPIE Microtechnologies. Bellingham, WA, USA: SPIE; 2017

[40] Snyder GF, Toberer ES. Complex thermoelectric materials. Nature Materials. 2008;7:105-114

[41] Andrew Truitt S. Nima Mahmoodi, A review on active wind energy harvesting designs. International Journal of Precision Engineering and Manufacturing. 2013;**14**:1667-1675

[42] Ma X, Zhou S. A review of flow-induced vibration energy harvesters. Energy Conversion and Management. 2022;**254**:115223

[43] Yazid MAM et al. Towards the implementation of energy harvesting for IoT sensor nodes in an early warning flood detection system. Journal of Communications. 2020;**15**:5

[44] https://www.uclahealth.org/u-magazine/

[45] https://e-peas.com/energy-harvesting/

[46] Tan YK, Panda SK. Optimized wind energy harvesting system using resistance emulator and active rectifier for wireless sensor nodes. IEEE Transactions on Power Electronics. 2011;**26**(1):38-50

[47] Cheng KWE. Energy management system for mobility and smart city. In: 2016 International Symposium on Electrical Engineering (ISEE). 2016. pp. 1-6

[48] Ram SK, Das BB, Mahapatra K, Mohanty SP, Choppali U. Energy perspectives in IoT driven smart villages and smart cities. IEEE Consumer Electronics Magazine. 2021;**10**(3):19-28

[49] Yu B. Design and experimental results of battery charging system for microgrid system. International Journal of Photoenergy. 2016;**2016**:1-6

[50] Chowdhury AH. Design Strategy for an Off-grid Solar-wind Hybrid Power System. Bangladesh: Dept. Elect. Electron. Eng; 2014

[51] Panhwar IH et al. Mitigating power fluctuations for energy storage in wind energy conversion system using supercapacitors. IEEE Access. 2020;**8**: 189747-189760. DOI: 10.1109/ACCESS.2020.3031446

[52] Kollimalla SK, Mishra MK, Ukil A, Gooi HB. Dc grid voltage regulation using new hess control strategy. IEEE Transactions on Sustainable Energy. 2017;**8**(2):772-781

[53] Yang H. A review of supercapacitor-based energy storage systems for microgrid applications. In: 2018 IEEE Power & Energy Society General Meeting (PESGM). 2018. pp. 1-5

[54] Novinc Ž. Kakvoća električne energije, Drugo izdanje. Zagreb: EDZ; 2006

[55] https://standards.globalspec.com/std/13493775/EN%2050160

[56] Mufti M, Lone SA, Iqbal SJ, Ahmad M, Ismail M. Super-capacitor based energy storage system for improved load frequency control. Electric Power Systems Research. 2009;**79**:226-233

[57] Denoun H, Zaouia M, Tamalouzt S, Bouheraoua M, Benamrouche N, Rekioua T, et al. Characterization and control of supercapacitors bank for stand-alone photovoltaic energy. Energy Procedia. 2013;**42**:539-548

[58] Miñambres-Marcos V, Romero-Cadaval E, et al. Power injection system for photovoltaic plants based on a multiconverter topology with DC-link capacitor voltage balancing. In: 12th International Conference on Optimization of Electrical and Electronic Equipment. 2010

[59] Tibude V, Tarnekar SG. Super capacitor for electric vehicle. International Journal of Engineering Research in Electrical and Electronic Engineering (IJEREEE). 2016;**2**:3

[60] Rajčić-Vujasinović MM, Stević ZM, Stanković ZD. Superkondenzatori. Hemijski pregled. 2002;**5**:108-112

[61] Stevic Z, Radovanovic I. Energy efficiency of electric vehicles. In: Stevic Z, editor. New Generation of Electric Vehicles [Internet]. London: IntechOpen; 2012. Available from: https://www.intechopen.com/chapters/41416. DOI: 10.5772/55237 [Cited: 04 August 2022]

[62] Bjekić M, Stević Z, Milovanović A, Antić S. Regulacija elektromotornih pogona. Čačak: Tehnički fakultet; 2010

[63] Labady A. Traction Brake Energy Regeneration By Supercapacitor Energy Storage System. Rotterdam: Eaton; 2019

[64] Dixon J, Ortuzar M. Ultracapacitors + DC-DC Converters in Regenerative Braking System. AESS System Magazine; August 2002

[65] Thong NA. Application of Supercapacitor in Electrical Energy Storage System. Singapore: National Universiti of Singapore; 2011

[66] Global Supercapacitor market [Internet]. 2022. Available from: https://www.maximizemarketresearch.com/market-report/global-SC-market/27055/. [Accessed: 2022-08-04]

[67] Raghavendra KVG, Vinoth R, Zeb K, Muralee Gopi CVV, Sambasivam S, Kummara MR, et al. An intuitive review of supercapacitors with recent progress and novel device applications. Journal of Energy Storage. 2020;**31**:101652

Chapter 2

Carbon Nanomaterials Based Supercapacitors: Recent Trends

Mohamed M. Atta and Rania M. Ahmed

Abstract

The increasing demand for renewable energy sources worldwide and the predicted depletion of current fossil fuel sources need continuous energy storage and conversion technology development. The use of supercapacitors (SC) as electrical energy storage devices in consumer electronics items and alternative power sources is an interesting and potentially lucrative area of application. Therefore, continuous developments are conducted to improve SC performance using different composites and nanocomposites. Carbon materials in SC are among the most important uses of this material. This chapter provides a short communication on recent progress in supercapacitor-based carbon materials. Various fundamental carbon allotropes were presented and debated, including fullerene, carbon nanotubes, and graphene-based supercapacitors.

Keywords: supercapacitor, graphene, carbon nanotubes, carbon, fullerene

1. Introduction

The increasing global energy demand due to a modern technology-dependent lifestyle puts increasing pressure on traditional nonrenewable energy resources such as fossil fuels [1, 2]. The global trend to use portable, flexible electronics and the increase in global environmental awareness push toward the use and continuous development of eco-friendly, sustainable energy conversion and storage devices [3, 4]. Supercapacitors are essential devices among energy storage devices because of their quick charging and discharging processes, high power densities, extended cycle lives, minimal maintenance requirements, long lifespans, and environmental friendliness [4]. Both the electric double layer capacitor, which stores energy by electrostatic means, and the pseudocapacitor, which keeps energy through redox reactions, are subtypes of the supercapacitor. In most cases, the hybrid electrodes can store energy in electrochemical and electrostatic methods [5]. Similar to all other devices, the performance of supercapacitors is highly dependent on the characteristics of the materials they utilize. Continuous developments of carbon materials are rapidly being employed in energy storage devices because of their advantages, including the simplicity of modifying and manipulating pore structure, surface functionality, surface area, and low cost. Furthermore, the structural integrity of the carbon framework ensures that the electrode material's cyclic stability and capacitance retention are maintained throughout time [6].

IntechOpen

In recent years, significant progress has been made in developing carbon nanostructure composites for high-performance energy conversion and storage devices. The recent advancement of nanoscience and nanotechnology has created novel graphitic carbon nanomaterials with multi-dimensions, such as two-dimensional (2D) graphene, (1D) carbon nanotubes (CNT), and dimensionless (0D) fullerene [7]. This chapter introduces recent supercapacitor advancements based on fundamental carbon nanostructures; graphene, carbon nanotubes, and fullerene.

2. Graphene

Graphene is an emerging carbon nanomaterial with an ideal 2D structure and unique electronic properties. On the other hand, the word graphene wasn't coined until 1986. Graphene is a single layer (2D) honeycomb-arranged carbon atom connected with sp^2 bonds. Graphene serves as the fundamental building block for the structure of all other carbon allotropes. Geim and Novoselov made the groundbreaking discovery in 2004 that single-layer and two to three-layer graphene nanosheets can stably survive in the environment [8, 9]. The exceptional qualities of graphene include its high electrical conductivity, high thermal conductivity (5000 W m^{-1} K^{-1}), high intrinsic charge mobility (250,000 cm^2 V^{-1} s^{-1}), and very high surface area (2630 m^2 g^{-1}), and high Young's modulus. Graphene also has a very high surface area (1.0 TPa).

Due to its low mass density, extremely large surface area, great conductivity, and high flexibility, graphene has widespread use in various sectors, including energy storage and conversion, electronic devices, sensors, polymer additives, and biological applications [10–15]. Different graphene synthesis approaches are reported as mechanical filiation, epitaxial growth, chemical vapor deposition, and reduction of graphene oxide [9]. The use of graphene and graphene composites as supercapacitor materials was the subject of numerous publications. The following are examples of recent research on graphene-based supercapacitors.

A high-performance supercapacitor was prepared based on a composite of carbonized wood cell chamber-reduced graphene oxide@PVA (CWCC-rGO@PVA) [16]. CWCC-rGO@PVA revealed a high specific capacitance of 288F g^{-1}, capacitance retention of 91%, energy density of 36 Wh kg^{-1}, and power density of 3600 W kg^{-1}.

The Co_3O_4/CoO nanoparticles were attached to reduced graphene oxide (rGO) nanosheets by microwave irradiation. The rGO@Co_3O4/CoO electrode showed excellent electrochemical performance of specific capacitance of 276.1 F g^{-1} and 82.37% capacitance retention after 10,000 cycles [17].

As supercapacitor electrodes, 3D flower-like spheres of $NiCo_2S_4$@Ni-Mo layered double hydroxide (LDH) nanocomposites grown in situ on reduced graphene oxide (RGO) were developed using a simple hydrothermal method [18]. For comparison, RGO@NiCo2S4 and RGO@NiMo-LDH electrodes were also prepared. The redox peaks in CV curves for the RGO@NiCo2S4@NiMo-LDH electrode were symmetric and had identical profiles as the scanning rate increased, demonstrating excellent pseudocapacitance behavior and rate capacity of the electrode. Charge–discharge curve platforms were more pronounced at varying current densities, suggesting the presence of a Faraday redox reaction. Capacity retention was very good for the RGO@$NiCo_2S_4$@NiMo-LDH electrode, with specific capacitances of 1346, 1336, 1305, 1294, 1283, and 1272 F g^{-1}at 1, 2, 4, 6, 8, and 10 A g^{-1}, respectively. The rated capacity of RGO@$NiCo_2S_4$@NiMo-LDH was higher than that of RGO@$NiCo_2S_4$.

$RGO@NiCo_2S_4@NiMo$-specific LDH's capacitance was greater than that of RGO@ $NiCo_2S4$ and RGO@NiMo-LDH taken separately, suggesting that the presence of several $NiCo_2S4$ nanosheets on graphene sheets may give more growth spots for NiMo-LDH nanosheets than a smooth graphene skeleton. It can be seen from the symmetrical charge–discharge curve that it has good electrochemical reversibility. The device can obtain a maximum energy density of 59.38 Wh kg^{-1} at a power density of 808.19 W kg^{-1} and maintain an energy density of 25.24 Wh kg^{-1} at a high power density of 8055.32 W kg^{-1}. The capacitance of the $RGO@NiCo_2S4@NiMo$-LDH electrode retained 80% of its initial capacitance after 10,000 cycles.

In another paper, graphene/MnV_2O_6 nanocomposite was prepared using solvothermal and liquid phase exfoliation processes. A maximum specific capacitance of 348 Fg^{-1} and capacitance retention of 88% was achieved after 3000 cycles for an optimal graphene/manganese vanadate ratio (1:8) sample [19].

A hybrid 2D platform was constructed from polypyrrole (PPy) /rGO and nickel-tungsten metal oxides. The prepared electrode showed excellent specific capacitance of 597 $F.g^{-1}$ with capacitance retention of 98.2% after 5000 cycles. The two-electrode device using the same electrode platform showed a specific capacitance of 361 $F.g^{-1}$ [20].

New hierarchical porous hybrid architecture consists of biomass-based porous carbon derived from Ganoderma lucidum residues (DDLG)/graphene composite aerogel were synthesized by chemical self-assembly and Vitamin C as a reducing agent [21]. Composites with 2.1, 3:1, 4:1, and 8:1 porous carbon ratios to GO were prepared. The large interconnected pores of DDLG were confirmed from SEM images. In addition, graphene aerogel retains the conventional three-dimensional network structure, and the sheet-like form of graphene is orientated unpredictably. Porous carbon/graphene composites feature a new three-dimensional hierarchical porous structure when the ratio of porous carbon to graphene is between 1:1 and 3:1. This ratio creates a densely packed structure. This is due to the graphene oxide sheet reduction process to conductive reduced graphene oxide resulting in forming a porous three-dimensional network structure around the BPC. When the ratio of porous carbon to graphene is exactly one to one, a system of porous carbon and graphene tightly packed together is produced.

Furthermore, the graphene self-assembled aerogel's structure dominates throughout the self-assembly process since just a few porous carbons are exposed owing to the high graphene concentration, and graphene nanoflakes cover the porous carbon. When the porous carbon to graphene ratio approaches 4:1, there are still two different types of porous structures, and the pore structure of porous carbon becomes more visible as the percentage increases. Furthermore, when the mass ratio of porous carbon increases to 8:1, graphene is shown to be distributed evenly throughout the porous carbon.

EIS measurements were used to analyze and compare the resistance characteristics of DDLGC and DDLGC/GO8. The DDLGC and DDLGC/GO8 ESRs were 0.53 and 0.46, demonstrating that the graphene-enhanced composite aerogel had significantly improved conductivity. The DDLGC/GO8-based electrode has a lower interfacial charge transfer resistance since the semicircle has a smaller diameter. A vertical line indicates capacitive behavior near to ideal [22]. A virtually vertical line was seen in the low-frequency region, suggesting high charge storage, rapid ion transport/diffusion, and excellent electrical double layer capacitor (EDLC) properties.

CV curves of DDLGC/GO8 at 5–100 mV s^{-1} show the creation of EDLC with rectangular curve shapes. The electrode of DDLGC/GO8 exhibited isosceles triangle shapes in the GCD plots at different current densities, demonstrating that the material's energy storage mechanism is a double-layer storage energy mechanism with good electrochemical reversibility. The specific capacitances of DDLGC and DDLGC/GO8 at different current densities were calculated. At a current density of 1 A g^{-1}, the specific capacitances of DDLGC and DDLGC/GO8 were determined to be 365.6 F g^{-1} and 366 F g^{-1}, respectively. DDLGC/GO8 has a substantially greater rate capacity at high current density than DDLGC, which may be attributed to the material's increased electron transfer efficiency at high scan rates [23] and the bigger average pore size and higher effective surface area. These findings show that adding graphene, another carbon element, may greatly enhance the capacitance characteristics of biomass-based porous materials.

3. Carbon nanotubes

In addition to their high electrical conductivity [24], unique pore structures, and improved power density in supercapacitors, carbon nanotubes (CNTs) have outstanding thermal stability, superior mechanical qualities, and unique pore structures. Powders made from commercially available CNTs are frequently used as collectors, either in conjunction with other pseudocapacitive materials or on their own, or as pseudocapacitive electrode materials [25]. Van der Waals force is the mechanism that allows CNTs in the electrode to link to one another. This increases the electrode's resistance leading to self-discharge as a consequence of poor adhesion. To address these risks, CNTs are grown on collectors, which can take the form of carbon cloth, graphene, stainless steel mesh, or nickel foam [26]. The following are some instances of research on the significance of CNTs in supercapacitor production.

Nitrogen-doped multiwalled carbon nanotubes (N-MWCNT) and carboxymethylcellulose (CMC) were combined by a hydrothermal process [27]. An N-MWCNT/CMC composite had an ultrasonication-mediated solvothermal reaction to produce the material. The good electrochemical characteristics and rapid redox reaction of the composite electrode in the presence of the PVA/H_2SO_4 gel electrolyte are deduced from the approximately rectangular shape of the cyclic voltammetry (CV) curves. The N-MWCNT/CMC composite electrode displayed a more significant current than the pure N-MWCNT, demonstrating its superior electrochemical performance and the crucial role of the CMC matrix inclusion on the CNTs in enhancing the electrode's capacitance. According to the galvanostatic charge–discharge (GCD) cyclic stability analysis performed for up to 4000 cycles at a scan rate of 2 Ag^{-1}, the N-MWCNT/CMC nanohybrid composite retained 96% of its initial capacity. The charge transfer (Rct) of the electrodes during the first and the one-thousandth cycles, as determined by the Nyquist plots, is approximately 0.9 and 35, respectively. The steady electrochemical characteristics are influenced as a result of this factor. At low frequencies, it was noticed that the phase angle for the impedance plot of the composite electrodes was greater than 45 degrees; this indicates that the composite electrodes have electrochemical capacitive capabilities.

The closed tips and fewer active sites of CNTs can limit their electrochemical performance. Therefore, Zhang and Xie [28] investigated a successful trial to open the tips of CNTs with oxygen and nitrogen functional groups by an effective chemical acid-etching method. The chemical vapor deposition (CVD) technique was used to perform the acidic treatment on the CNTs fabricated. Li^+-based electrolyte provided the best electrochemical performance of the functionalized and tip-open CNTs (FTO-CNTs) compared to the other investigated types of ions as Na^+ and Mg^{2+}.

The areal capacitance obtained from GCD curves for the FTO-CNTs indicates improved electrochemical performance. Compared to CNT growth on carbon cloth (CCC) and carbon cloth (CC), FTO-CNTs have the highest areal capacitance due to their largest CV area as determined by CV curves at a scan rate of 20 mV/s. The functionalization and tip-opening of the CNTs may explain the higher capacitance of FTO-CNTs compared to that of CCC. The higher number of oxidation–reduction reactions is responsible for the greatest charge–charge transfer resistance between ions and electrons (R_{ct}) in FTO–CNTs. The movement of ions from the open tip into the interior of the CNTs may be responsible for the higher diffusion resistances (σ) exhibited by FTO-CNTs compared to those of CCC. Since there were more entrance locations for the diffusion of ions in the FTO-CNTs due to their open tips, a scan rate of 10 mV induced a greater diffusion-controlled capacitance (75%) higher than CCC (65%).

Yang et al. developed an innovative method for dealing with polymer waste and high-value-added recycling of resources [29]. In this study, the researchers investigated a great success in treating polypropylene face mask wastes, a source of environmental pollution, to be useful by carbonizing them into CNTs. Yang et al. proposed employing the manufactured waste face mask CNTs as electrode material in supercapacitors to achieve extra financial benefits. The CNTs were produced using Ni–Fe bimetallic catalysts with varying molar ratios NiFeX (X = 1 to 5 and NiFe/Al = 1).

CV curves of CNT samples appeared in approximately rectangular patterns with broad redox peaks. The development of broad redox peaks explained by the insertion of functional groups comprising nitrogen and oxygen on the surface of carbon nanotubes. The best value of the ratio capacitance was detected in the CNT-$NiFe_3$ sample. For the CNTs sample, the electric double-layer capacitance features were proved by results obtained from capacitance performance (CP) curves which provided an isosceles-like triangle at a current density of 1 A/g and a range from −0.8 V to zero V. After 10,000 cycles, CNT-$NiFe_3$ electrodes have high cycling stability with capacitance retention of 85.41% from the initial value. Also, within a current density of 1 Ag^{-1}, they attain a specific capacitance of 56.04 F/g. Due to the bamboo-like shape of the carbon nanotubes, CNT-$NiFe_3$ can be purified to achieve the maximum specific surface area and N-doped concentration.

A green, simple processing protocol proposed by Bathula et al. [30] utilizing mechanochemical grinding to synthesize hybrid nanostructures of cobalt oxide on nitrogen-doped multiwalled carbon nanotubes (Co3O4-NMWCNT). The NMWCNT in its original form exhibited wire-like geomorphology; however, Co_3O_4 consists of clusters of pieces, and the NMWCNT-Co_3O_4 composite includes an interconnected tube structure. The electrochemical properties of symmetric devices made with NMWCNT, and NMWCNT-Co_3O_4 electrodes were studied. CV curves of both electrodes verified the EDLC behavior and Faradic reaction, respectively. The enclosed area of the CV of the NMWCNT-Co_3O_4 device is nearly twice that of the NMWCNT device, indicating that the Co3O4 and NMWCNT have a synergy effect. Both materials have remarkable rapid charging and discharging potential. Random CV curves illustrated that the form of CV curves for Co_3O_4-NMWCNT was maintained across all cycles (indicating exceptional structural stability).

In another article, a composite of polypyrrolopyrrolethieno thiophene (PDPT) and carbon nanotube (CNT) was created by Bathula et al. [31] to test its viability as a hybrid electrode material. The structure of PDPT is based on DPP (π-conjugated polymer), which includes moieties of both sulfur and nitrogen heterocyclic. DPT accumulates a donor-acceptor (D–A) interface utilizing chemical exfoliation suggested for electron-accepting bulk. Ultrasonic vibrations caused exfoliation in this particular investigation. To obtain an intermolecular hydrogen connection and necessary D–A and p–p packing, the authors investigated a successful mixture between CNTs and bulk DPT nanofibers. Afterwards, a standardized PDPT-CNT composite suspension was produced from the accumulation of the insoluble DPT. The GCD results showed the specific capacitance of PDPT-CNT and PDPT are 126, 90, 60, 30, and 10; and 42, 26, 16, 12, and 5 F/g, detected at current densities of 0.5, 1, 2, 3, and 5 A/g, respectively. Moreover, at a power density of 450 W/kg, the PDPT-CNT device has a maximum energy density of 15.7 W.h/kg.

Zhang et al. [32] constructed a novel wire-shaped coaxial supercapacitor with exceptional performance, made of carbon wires (CW)@MnO_2/PVA-KOH/carbon nanotubes (CNTs). For the inner electrode, copper wire was utilized as a current collector to solve the problem of the low electric conductivity of MnO_2. However, carbon nanotubes generated via in-situ chemical vapor deposition (CVD) served as the outer electrode, with cobalt-based catalyst particles uniformly dispersed across the surface of SiO_2. Then, the device was created by removing the SiO_2 layer and filling it with a polyvinyl alcohol-KOH (PVA-KOH) gel electrolyte, simultaneously using the hydrothermal method. At a power density of 37 mW cm^{-3}, the wire-shaped supercapacitor had the highest volumetric energy density of 0.16 mWh cm^{-3}, dropping to 0.12 mWh cm^{-3} at 62.6 mW cm^{-3} power density. The asymmetrical quasi-rectangle shape of a wire-shaped supercapacitor obtained from CV curves indicates its exceptional electrochemical performance. The observed semicircle in the Nyquist curve of electrochemical impedance spectroscopy (EIS) for the wire-shaped supercapacitor at high frequencies describes the resistance of the charge transfer, which has a value of 1815 W. The equivalent internal resistance value was 22.79 W, deduced from the semicircle's intercept with the real axis at high frequencies. The capacitance retention continued high and stable; after 4000 cycles of charging and discharging, the capacitance of the wire-shaped supercapacitor exhibits excellent retention of over 90.38%. From the Ragone plot, at a power density of 37 mW cm^3, the wire-shaped supercapacitor has the greatest volumetric energy density (0.16 mWh cm^3). More significantly, it may continue to be 0.12 mWh cm^3 even when the power density reaches 62.6 mW cm^3.

4. Fullerene

Fullerenes or Buckyballs are a novel class of carbon nanomaterials. The basic elements of fullerenes are called isomers. Their homologs range from the lower homologs that have received the most attention, such as C60 and C70, to higher fullerenes, such as C240, C540, and C720. They have shown to be valuable in various scientific domains, including separating and identifying different chemical species. Fullerene was first produced by Kroto, Curl, and Smalley via laser-induced evaporation of graphite. As a result, the discovery of Buckminsterfullerene, also known as C60, resulted from a research study that connected synthetic chemistry, microwave spectroscopy, and radio-astronomy. Fullerene was born from the search to reproduce poly acetylenes discovered in interstellar space [33].

Fullerenes have a structure composed of sp2 carbons with distinct chemical and physical characteristics and a highly symmetrical cage with varying widths

(C60, C76, etc.) [34]. Thanks to their excellent electrochemical stability, small size, unique shape, and well-ordered structure [35], they enable their use in energy conversion systems.

Thanks to their excellent electrochemical stability, small size, unique shape, and well-ordered structure, they enable their use in energy conversion systems. Fullerenes' distinctive 0D structure makes them valuable building blocks for supramolecular assemblies and micro/nano functional materials used in drug delivery [36], photovoltaic devices [37], optoelectronics [38], sensors [39], catalysis [40], and other fields.

The fullerene molecule C60, which has a structure of 60 carbon atoms, 12 pentagonal C5-C5 single bonds, and C5 = C6 double bonds (20 hexagons), is the most often employed in supercapacitors. The following are some studies on the importance of fullerene in supercapacitor manufacture.

Activated fullerene (A-C60) decorated over zinc cobaltite (A-C60-ZCO) has been synthesized by a solvothermal approach as a supercapacitor electrode [41]. The greater enclosed area of the CV and the well-defined redox peaks suggest that A-C60-ZCO has a high specific capacitance and a strong pseudo capacitive nature. It was reported that the specific capacitance value is better for the 10 wt.% of A-C60 in ZCO loading than for the 2, 5, and 15 wt.% loadings. So, a composite with 10 wt.% A-C60 loading is the best for further electrochemical studies [41]. The character of the CV curve of A-C60-ZCO stays the same, except for a shift in peak position even at a higher scan rate (100 mV/s), indicating that the as manufactured material possesses rapid and reversible faradic performance. At scan speeds of 1, 5, 10, 20, 40, 50, 70, and 100 mV/sec, the A-C60-ZCO has volumetric specific capacitances of 593.2, 554.18, 506.2, 412.3, 332.7, 296.584, 260.696, 221.76 F/g. With an increase in scan rate, specific capacitance decreases as internal resistance becomes more dominant. The pseudocapacitive character of the active material is firmly confirmed by all GCD curves, resembling separate plateau areas compared to CV curves. Because of the partial ion migration toward the core of the active material, which may be controlled by limiting the loading quantity of active material, the specific capacitance value at higher current densities shows a small decline.

Under the same current density, -C60-ZCO has the longest charge/discharge time among ZCO, A-C60, and C60, indicating that A-C60-ZCO has the highest specific capacitance. At a current density of 2 A/g, the specific capacitance of A-C60-ZCO, ZCO, A-C60, and C60 electrodes is determined to be 269.81, 124.05, 34.41, and 24.18 F/g, respectively. The synergistic impact of the pseudo capacitive ZCO and A-C60 increases specific capacitance. All of the GCD curves have plateaus, which is strong evidence that the active material is pseudo capacitive and consistent with CV curves. The calculated specific capacitance values were 269.81, 144.36, 106.53, 84.06, 33.03, 27.89, 23.11, and 19.56 F/g at a current density of 2, 3, 4, 5, 7, 8, 9, and 10 F/g respectively.

In another work, a composite from polyaniline (PANI)/fullerene derivative (PCBM) Phenyl-C60-butyric acid methyl ester was constructed and tested as supercapacitor materials [42]. By varying the ratios of PCBM, different PANI/ PCBMx (where x = 0, 2.5, 5, and 10) were prepared. It was concluded that the PANI/PCBM electrodes had a higher specific capacitance than PANI due to the synergetic effect of PANI and PCBM. Also, it was found that the PANI/PCBM5 had the highest specific capacitance of 2609 F/g compared to 1216, 1882, and 1770 F/g for pure PANI, PANI/PCBM2.5, and PANI/PCBM10. The decreasing of specific capacitance of nanocomposite electrodes with PCBM content higher than 5 wt.% is ascribed to a larger size of PCBM, which decreases surface area.

3D pore structure produced C60 molecules into graphene sheets by hydrothermal approach to enhance their electrochemical performance [42]. The CV curves of mC60/graphene composite revealed the EDLC and pseudocapacitors. The electrochemical dependence on mass ratio, temperature, and reaction time was studied. It was found that typically when the mass ratio of C60 to GO is 1:8, reaction time is 12 hr., and temperature is 150oC, the specific capacitance reaches 332.3F/g compared to 215.1 F/g for pure reduced graphene oxide. It was concluded from GCD curves that the mass ratio of C60 to GO is 1:8 is the best for optimizing the composite charge/discharge performance. C60 molecules into graphene sheets by hydrothermal approach to enhance their electrochemical performance [43]. The CV curves of mC60/graphene composite revealed the EDLC and pseudocapacitors. The electrochemical dependence on mass ratio, temperature, and reaction time was studied. It was found that typically when the mass ratio of C60 to GO is 1:8, reaction time is 12 hr., and temperature is 150°C, the specific capacitance reaches 332.3F/g compared to 215.1 F/g for pure reduced graphene oxide. It was concluded from GCD curves that the mass ratio of C60 to GO is 1:8 is the best for optimizing the composite charge/discharge performance.

A novel supercapacitor electrode was created using a carbon nano-onion(multilayer fullerene) /manganese dioxide/iron oxide ($CNO/MnO_2/Fe_3O_4$) nanocomposite [44]. The electrochemical performance of prepared supercapacitors composed of MnO_2, CNO, MnO_2/Fe_3O_4, and $CNO/MnO_2/Fe_3O_4$ nanocomposite was investigated. The rectangular shapes of CV curves of electrodes were established. The rise in super-capacitance of the $CNO/MnO_2/Fe_3O_4$ electrode is due to the increased surface area of the CNO and the presence of MnO_2 and Fe_3O_4, which increases the adsorption/desorption of cation and onion on the nanocomposite surface. The super-capacitive current of the $CNO/MnO_2/Fe_3O_4$ was higher than metal oxides electrodes due to the presence of CNO with a high surface area. It was observed that the Metal oxide electrodes have less symmetry than those containing CNO. Furthermore, The $CNO/MnO_2/Fe_3O_4$ nanocomposite electrode's longer discharge duration implies improved electrode quality. The calculated specific capacitance of $CNO/MnO_2/Fe_3O_4$ electrodes was higher than other electrodes. At 1, 2, 3, and 4 A/g, $CNO/MnO_2/Fe_3O_4$ had a specific capacitance of 1130, 972.50, 900, and 730 F/g, while MnO_2/Fe_3O_4 had 571.25, 537.50, 442.50, and 400 F/g. Specific capacitances were 487.5 F/g at 1 A/g, 415 F/g at 2 A/g, 375 F/g at 3 A/g, and 340 F/g at 4 A/g for CNO. MnO_2's capacitance at 1 to 4 A/g was 382.94, 326.14, 285.12, and 202.59 [44].

Using the stacking interactions of graphene with aromatic rings of functionalized fullerenes created and produced several unique graphene-based nanomaterials. To assure strong contacts and stable assembly of fullerenes on the surface of graphene, C60, C70, and Sc3N@C80 fullerene derivatives containing biphenyl, naphthalene, phenanthrene, or pyrene moieties were produced [45]. Graphene coated with bis-naphthalene C_{70} fullerene malonate (G-BN7) revealed a 15% higher capacitance than graphene before modification, with a specific capacitance value of 56.15 F/g. Thus, naphthalene is the most suitable substitution for introducing fullerene derivatives on the graphene surface via π–π stacking. Additionally, compared to C60 and Sc3N@C_{80}, the C_{70} fullerene core delivered the greatest results.

The low long-range conductivity of fullerene severely hinders the performance of supercapacitors that use this material. It is therefore anticipated that active carbons based on fullerene will have large capacitances when they are developed. By manipulating fullerene self-assembly with a cobalt tetramethoxy phenylporphyrin (CoTMPP) and pyrolysis, mesoporous carbon composites doped with varying concentrations of cobalt (Co) and nitrogen (N) were synthesized by Jiang et al. [46]. C_{60} crystals encapsulated CoTMPP, which underwent carbonization to become actively-bound Co–N in the carbon structures. The ratio of CoTMPP in C_{60} crystals and the distribution state in superstructures influence the concentration of Co–N. The electrochemical performance of porous carbon composite was greatly improved by Co–N. The fabricated carbon composite demonstrated an improved specific capacitance of 416.31 F g^{-1} at 1 A g^{-1}, which is over ten times greater than that of the pristine C60, and had no activity loss after at least 5000 cycles.

Orderly mesoporous fullerene/carbon hybrids were synthesized by combining the fullerene precursor in chloronaphthalene with varying quantities of sucrose

Electrode material	Specific/Volumetric capacitance	Cyclic stability	Ref
$ZnFe_2O_4$- RGO	1419 F/g	93% retention after 5000 cycles	[48]
Poly (3-hexyl-thiophene-2, 5-diyl)/ CNT	245.8 F/g	80.5% retention after 1000 cycles	[49]
Graphene/ MoS_2	290 F/ cm^3	90% retention after 10,000 cycles	[50]
PANI/CNT	541 F/ g	90% retention after 25 cycles	[35]
PANI/fullerene	2201 F/g	96% retention after 1000 cycles.	[42]
Ag/Bi nanoparticle anchored CNT	1372F/g	101.3% retention after 10,000 cycles	[51]
MnO_2@CNT	386 F/ g	93.6% retention after 5000 cycles	[52]
MoS_2/Mn- metal organic frameworks (MOF)/CNT	862.73F /g	71.4% retention after 5000 cycles	[53]
CeO_2/graphene	782 F/g	82% retention after 6000 cycles	[54]
nitrogen-doped carbon nano-onions (N-CNO)	205 F/ g	96% retention after 5000 cycles	[55]
MXene/graphene	183.5 F/ cm^3	75% retention after 3000 cycles	[56]
CNT@PANI	138F/g	86% retention after 1000 cycles	[57]
RGO/MWCNT/ZrO2	357 F/ g	98% retention after 5000 cycles	[58]
MnO_2/$CoWO_4$/ nitrogen-doped carbon nanoonions (NCNO)	536 F/g	96% retention after 3000 cycles	[59]
RuO_2 quantum dots / RGO	1120 F/ g	89% retention after 10,000 cycles	[60]
Tetraaniline(TA)/porous RGO	85.6 F/g	104.7% after 10,000 cycles	[61]

Table 1.
Some latest fabricated carbon-based supercapacitor electrodes.

and employing mesoporous silica SBA-15 as a template [47]. Different samples MC_{60}@C-X, where X denotes the weight ratio of the fullerene C_{60} and sucrose were prepared. The ideal EDLC behavior was observed for all samples as deduced from CV curves. By decreasing the C60/sucrose ratio from 2 to 1.33, the calculated specific capacitance increased and then reduced as the ratio fell to 0.8. The highest specific capacitance of 213 F/g at 0.5 A/g was achieved for the MC_{60}@C-1.33 electrode, which is higher than pure mesoporous fullerene prepared without sucrose molecules. Using superior textural parameters, the authors of this study demonstrated that incorporating carbon into the fullerene matrix enhanced the electrical transport and diffusion of the electrolytes. In addition, the research findings suggested that the presence of carbon layers between the fullerenes helped to strengthen the connection between the molecules of fullerene and promoted the electronic transition.

Table 1 summarizes some features of carbon-based supercapacitor electrodes that have been recently reported.

Author details

Mohamed M. Atta[1*] and Rania M. Ahmed[2]

1 Radiation Physics Dept. National Center for Radiation Research and Technology (NCRRT), Egyptian Atomic Energy Authority, Cairo, Egypt

2 Faculty of Science, Physics Department, Zagazig University, Zagazig, Egypt

*Address all correspondence to: mohatta2014@gmail.com

IntechOpen

© 2022 The Author(s). Licensee IntechOpen. This chapter is distributed under the terms of the Creative Commons Attribution License (http://creativecommons.org/licenses/by/3.0), which permits unrestricted use, distribution, and reproduction in any medium, provided the original work is properly cited.

References

[1] Atta M, Fahim RA. Flexible and wearable supercapacitors: A short review. Journal of Energy Storage. 2021;**44**:103475

[2] Dhakal G, Mohapatra D, Tamang TL, Lee M, Lee YR, Shim J-J. Redox-additive electrolyte–driven enhancement of the electrochemical energy storage performance of asymmetric Co3O4// carbon nano-onions supercapacitors. Energy. 2021;**218**:119436

[3] Atta M, Abdel Maksoud M, Sallam O, Awed A. Gamma irradiation synthesis of wearable supercapacitor based on reduced graphene oxide/cotton yarn electrode. Journal of Materials Science: Materials in Electronics. 2021;**32**:3688-3698

[4] Banavath R, Nemala SS, Kim S-H, Bohm S, Ansari MZ, Mohapatra D, et al. Industrially scalable exfoliated graphene nanoplatelets by high-pressure airless spray technique for high-performance supercapacitors. FlatChem. 2022;**33**: 100373

[5] Lakra R, Kumar R, Sahoo PK, Thatoi D, Soam A. A mini-review: Graphene based composites for supercapacitor application. Inorganic Chemistry Communications. 2021;**133**:108929

[6] Rawat S, Mishra RK, Bhaskar T. Biomass derived functional carbon materials for supercapacitor applications. Chemosphere. 2022;**286**:131961

[7] Chen X, Paul R, Dai L. Carbon-based supercapacitors for efficient energy storage. National Science Review. 2017;**4**:453-489

[8] Inagaki M, Kang F. Materials Science and Engineering of Carbon: Fundamentals. USA: Butterworth-Heinemann; 2014

[9] Xiang Q, Yu J, Jaroniec M. Graphene-based semiconductor photocatalysts. Chemical Society Reviews. 2012;**41**:782-796

[10] Olabi AG, Abdelkareem MA, Wilberforce T, Sayed ET. Application of graphene in energy storage device–a review. Renewable and Sustainable Energy Reviews. 2021;**135**:110026

[11] Atta M, Ashry H, Nasr G, El-Rehim A. Electrical, thermal and electrochemical properties of γ-ray-reduced graphene oxide. International Journal of Minerals, Metallurgy and Materials. 2021;**28**:1726-1734

[12] Ahmed R, Atta M, Taha E. Optical spectroscopy, thermal analysis, and dynamic mechanical properties of graphene nano-platelets reinforced polyvinylchloride. Journal of Materials Science: Materials in Electronics. 2021;**32**:22699-22717

[13] Zhang H, He R, Niu Y, Han F, Li J, Zhang X, et al. Graphene-enabled wearable sensors for healthcare monitoring. Biosensors and Bioelectronics. 2022;**197**:113777

[14] Kodous AS, Atta M, Abdel-Hamid GR, Ashry H. Anti-metastatic cancer activity of ultrasonic synthesized reduced graphene oxide/ copper composites. Chemical Papers. 2022;**76**:373-384

[15] Atta M, Habieb M, Mohamed MAEH, Lotfy D, Taha EO. Radiation-assisted reduction of graphene oxide by aloe vera and ginger and their antioxidant and anti-inflammatory roles against male

mice liver injury induced by gamma radiation. New Journal of Chemistry. 2022;**46**:4406-4420

[16] Xiong C, Li B, Duan C, Dai L, Nie S, Qin C, et al. Carbonized wood cell chamber-reduced graphene oxide@ PVA flexible conductive material for supercapacitor, strain sensing and moisture-electric generation applications. Chemical Engineering Journal. 2021;**418**:129518

[17] Kumar R, Sahoo S, Tan WK, Kawamura G, Matsuda A, Kar KK. Microwave-assisted thin reduced graphene oxide-cobalt oxide nanoparticles as hybrids for electrode materials in supercapacitor. Journal of Energy Storage. 2021;**40**:102724

[18] Cheng C, Zou Y, Xu F, Xiang C, Sui Q, Zhang J, et al. Ultrathin graphene@ NiCo2S4@ Ni-Mo layered double hydroxide with a 3D hierarchical flowers structure as a high performance positive electrode for hybrid supercapacitor. Journal of Energy Storage. 2022;**52**:105049

[19] Low WH, Lim SS, Siong CW, Chia CH, Khiew PS. One dimensional MnV2O6 nanobelts on graphene as outstanding electrode material for high energy density symmetric supercapacitor. Ceramics International. 2021;**47**:9560-9568

[20] Hashemi SA, Mousavi SM, Naderi HR, Bahrani S, Arjmand M, Hagfeldt A, et al. Reinforced polypyrrole with 2D graphene flakes decorated with interconnected nickel-tungsten metal oxide complex toward superiorly stable supercapacitor. Chemical Engineering Journal. 2021;**418**:129396

[21] Xu M, Wang A, Xiang Y, Niu J. Biomass-based porous carbon/graphene self-assembled composite aerogels for high-rate performance supercapacitor. Journal of Cleaner Production. 2021;**315**:128110

[22] Yakaboylu GA, Yumak T, Jiang C, Zondlo JW, Wang J, Sabolsky EM. Preparation of highly porous carbon through slow oxidative torrefaction, pyrolysis, and chemical activation of lignocellulosic biomass for high-performance supercapacitors. Energy & Fuels. 2019;**33**:9309-9329

[23] Zhou H, Wu S, Wang H, Li Y, Liu X, Zhou Y. The preparation of porous carbon materials derived from bio-protic ionic liquid with application in flexible solid-state supercapacitors. Journal of Hazardous Materials. 2021;**402**:124023

[24] Najib S, Erdem E. Current progress achieved in novel materials for supercapacitor electrodes: mini review. Nanoscale Advances. 2019;**1**:2817-2827

[25] Lokhande V, Lee SJ, Lokhande A, Kim JH, Ji T. 1.5 V symmetric supercapacitor device based on hydrothermally synthesized carbon nanotubes and cobalt tungstate nanocomposite electrodes. Materials Chemistry and Physics. 2018;**211**:214-224

[26] Liang X, Zhao L, Wang Q, Ma Y, Zhang D. A dynamic stretchable and self-healable supercapacitor with a CNT/graphene/PANI composite film. Nanoscale. 2018;**10**:22329-22334

[27] Basivi PK, Ramesh S, Kakani V, Yadav H, Bathula C, Afsar N, et al. Ultrasonication-mediated nitrogen-doped multiwalled carbon nanotubes involving carboxy methylcellulose composite for solid-state supercapacitor applications. Scientific Reports. 2021;**11**:1-12

[28] Zhang Y, Xie E. Functionalized and tip-open carbon nanotubes

for high-performance symmetric supercapacitors. Dalton Transactions. 2021;**50**:12982-12989

[29] Yang W, Cao L, Li W, Du X, Lin Z, Zhang P. Carbon nanotube prepared by catalytic pyrolysis as the electrode for supercapacitors from polypropylene wasted face masks. Ionics. 2022;**28**:1-12

[30] Bathula C, Rabani I, Ramesh S, Lee S-H, Palem RR, Ahmed ATA, et al. Highly efficient solid-state synthesis of Co3O4 on multiwalled carbon nanotubes for supercapacitors. Journal of Alloys and Compounds. 2021;**887**:161307

[31] Bathula C, Rabani I, Kadam A, Opoku H, Patil SA, Shreshta NK, et al. Sonochemically exfoliated polymer-carbon nanotube interface for high performance supercapacitors. Journal of Colloid and Interface Science. 2022;**606**:1792-1799

[32] Zhang Q, Yang F, Zhang C, Dong H, Sui J, Yu L, et al. A novel wire-shaped supercapacitor based on MnO2 nanoflakes and carbon nanotubes with high performance synthesized by sacrificial template method. Applied Surface Science. 2021;**551**:149417

[33] Henson R. The history of carbon 60 or buckminsterfullerene. Archived from The Original on. 2013

[34] Kazemzadeh H, Mozafari M. Fullerene-based delivery systems. Drug Discovery Today. 2019;**24**:898-905

[35] Panasenko IV, Bulavskiy MO, Iurchenkova AA, Aguilar-Martinez Y, Fedorov FS, Fedorovskaya EO, et al. Flexible supercapacitors based on free-standing polyaniline/single-walled carbon nanotube films. Journal of Power Sources. 2022;**541**:231691

[36] Kumar M, Raza K. C60-fullerenes as drug delivery carriers for anticancer agents: Promises and hurdles. Pharmaceutical Nanotechnology. 2017;**5**:169-179

[37] Collavini S, Delgado JL. Fullerenes: The stars of photovoltaics, sustainable. Energy & Fuels. 2018;**2**:2480-2493

[38] Das S, Presselt M. Progress and development in structural and optoelectronic tunability of supramolecular nonbonded fullerene assemblies. Journal of Materials Chemistry C. 2019;**7**:6194-6216

[39] Shetti NP, Mishra A, Basu S, Aminabhavi TM. Versatile fullerenes as sensor materials. Materials Today Chemistry. 2021;**20**:100454

[40] Dorel R, Echavarren AM. From palladium to gold catalysis for the synthesis of crushed fullerenes and Acenes. Accounts of Chemical Research. 2019;**52**:1812-1823

[41] Mohanty R, Swain G, Parida K, Parida K. Enhanced electrochemical performance of flexible asymmetric supercapacitor based on novel nanostructured activated fullerene anchored zinc cobaltite. Journal of Alloys and Compounds. 2022;**919**:165753

[42] Ramadan A, Anas M, Ebrahim S, Soliman M, Abou-Aly A. Polyaniline/fullerene derivative nanocomposite for highly efficient supercapacitor electrode. International Journal of Hydrogen Energy. 2020;**45**:16254-16265

[43] Cheng L, Li X-J, Li J, Qiu H-X, Xue Y-H, Evgenyevna K-I, et al. Construction of three-dimensional all-carbon C60/graphene hybrids and their use in high performance supercapacitors. New Carbon Materials. 2020;**35**:684-695

[44] Sohouli E, Adib K, Maddah B, Najafi M. Preparation of a supercapacitor

electrode based on carbon nano-onions/manganese dioxide/iron oxide nanocomposite. Journal of Energy Storage. 2022;**52**:104987

[45] PiotrowskiP, FedorczykA, GrebowskiJ, Krogul-Sobczak A. Functionalization of graphene by π–π stacking with C60/C70/Sc3N@ C80 fullerene derivatives for supercapacitor electrode materials, C. Journal of Carbon Research. 2022;**8**:1-17

[46] Jiang B, Zhang G, Tang Q, Meng F, Zhou D, Zhao W, et al. Tailoring co-doping of cobalt and nitrogen in a fullerene-based carbon composite and its effect on the supercapacitive performance. Materials Advances. 2022;**3**:1539-1546

[47] Baskar AV, Ruban AM, Davidraj JM, Singh G, Al-Muhtaseb AAH, Lee JM, et al. Single-step synthesis of 2D mesoporous C60/carbon hybrids for supercapacitor and Li-ion battery applications. Bulletin of the Chemical Society of Japan. 2021;**94**:133-140

[48] Askari MB, Salarizadeh P, Seifi M, Di Bartolomeo A. ZnFe2O4 nanorods on reduced graphene oxide as advanced supercapacitor electrodes. Journal of Alloys and Compounds. 2021;**860**:158497

[49] Shokry A, Karim M, Khalil M, Ebrahim S, El Nady J. Supercapacitor based on polymeric binary composite of polythiophene and single-walled carbon nanotubes. Scientific Reports. 2022;**12**:1-13

[50] Zhuo Y, Prestat E, Kinloch IA, Bissett MA. Self-assembled 1T-MoS2/functionalized graphene composite electrodes for supercapacitor devices. ACS Applied Energy Materials. 2022;**5**:61-70

[51] Cha Y, Kim T, Seo B, Choi W. Combustion-driven synthesis route for bimetallic Ag–Bi nanoparticle-anchored carbon nanotube electrodes for high-performance supercapacitors. Carbon. 2022;**198**:11-21

[52] Teng S, Shi S, Wang G, Xiang Y, Wan G. Ozone-activated CNTs to induce uniform coating of MnO2 as high-performance supercapacitor electrodes. Fullerenes, Nanotubes, and Carbon Nanostructures. 2022:1-7

[53] Peng W, Song N, Su Z, Wang J, Chen K, Li S, et al. Two-dimensional MoS2/Mn-MOF/multi-walled carbon nanotubes composite material for high-performance supercapacitors. Microchemical Journal. 2022;**179**:107506

[54] Jayashree M, Sharmila V, Meganathan K, Boopathi Raja R, Parthibavarman M, Shkir M, et al. Design and fabrication of graphene anchored CeO2 hybrid nanocomposite electrodes for high performance energy storage device applications. Inorganic Chemistry Communications. 2021;**132**:108838

[55] Pallavolu MR, Gaddam N, Banerjee AN, Nallapureddy RR, Joo SW. Superior energy-power performance of N-doped carbon nano-onions-based asymmetric and symmetric supercapacitor devices. International Journal of Energy Research. 2022;**46**:1234-1249

[56] Wen D, Ying G, Liu L, Li Y, Sun C, Hu C, et al. Direct inkjet printing of flexible MXene/graphene composite films for supercapacitor electrodes. Journal of Alloys and Compounds. 2022;**900**:163436

[57] Wang H, Biswas SK, Zhu S, Lu Y, Yue Y, Han J, et al. Self-healable electro-conductive hydrogels based on core-shell structured nanocellulose/carbon nanotubes hybrids for use as

flexible supercapacitors. Nanomaterials. 2020;**10**:112

[58] Jose J, Vigneshwaran J, Baby A, Viswanathan R, Jose SP, Sreeja P. Dimensionally engineered ternary nanocomposite of reduced graphene oxide/multiwalled carbon nanotubes/zirconium oxide for supercapacitors. Journal of Alloys and Compounds. 2022;**896**:163067

[59] Sohouli E, Adib K, Maddah B, Najafi M. Manganese dioxide/cobalt tungstate/nitrogen-doped carbon nano-onions nanocomposite as new supercapacitor electrode. Ceramics International. 2022;**48**:295-303

[60] Zhao J, Zhang J, Yin H, Zhao Y, Xu G, Yuan J, et al. Ultra-fine ruthenium oxide quantum dots/reduced graphene oxide composite as electrodes for high-performance supercapacitors. Nanomaterials. 2022;**12**:1210

[61] Wang D, Liu P. Well-defined tetraaniline deposited graphene via mixed self-assembly for high-performance flexible supercapacitor application. Surfaces and Interfaces. 2022;**29**:101793

Chapter 3

Perspective Chapter: Graphene Based Nanocomposites for Supercapacitor Electrodes

Kefayat Ullah, Bakht Mand Khan, Amin Ur Rashid and Won Chun Oh

Abstract

The demand for engineering and advancement of supercapacitor electrodes are increasing globally. To address the production and storage capacity of the supercapacitor electrodes, the development of new kind of composite materials are highly needful. To design materials with high surface area, excellent conductivity, porosity, and mechanical stability are the main critical points that need to be addressed. Various strategies have been utilized to fabricate excellent composite materials for supercapacitor electrodes. The effect of many composite materials was found to enhance the cyclability and storage capacities of the supercapacitor electrodes. In a class of materials, graphene-based nanocomposites and their derivatives were found to be the most excellent and suitable candidates to design and fabricate supercapacitor electrodes. The alliance of several active materials when analyzed with graphene and its derivatives was found to improve further the performance and stability of supercapacitor electrodes.

Keywords: graphene, nanocomposites, supercapacitor, electrodes, metal oxides

1. Introduction

Supercapacitor is an attractive electrochemical device that fulfills the requirement of all advanced electronic and electrical devices [1]. Supercapacitors have gained a lot of attention and have been employed in various fields, including various electronic devices, power supplies, and electric vehicles due to their high-power densities, rapid charge/discharge rates, and exceptional cycling stability [2–5]. Supercapacitors are divided into two categories depending on their charge storage capacity: 1) electrical double-layer capacitors (EDLCs) made of various carbon-based materials, while 2) pseudocapacitors made of transition metal oxides and other conductive polymers as active materials [6–9]. Ions are arranged on the interface of the electrode and electrolytes in EDLC charge storage. In this mechanism, ions adsorb and desorb quickly at the electrolyte-electrode contact, resulting in high power density. Furthermore, no chemical reactions are involved in this charge storage process; simply the transport and adjustment of ions occur during the process [10]. The energy storage of pseudocapacitors, on the other hand, is caused by the fast redox reactions or faradaic mechanisms of the electrochemically active materials in the electrode [11]. When EDLCs and pseudocapacitors are merged into a single device, a new subcategory known as hybrid capacitors is constructed [12]. The most important components of supercapacitors are the electrode materials that are used in cathodes and anodes [13]. The essential component of a supercapacitor is the electrode material, which directly influences the electrochemical performance [14]. As a result, the development and application of innovative electrode materials are critical for improving supercapacitor performance.

Graphene-based composite electrodes are regarded as one of the most effective electrode materials because of their impressive chemical, mechanical, and physical properties, such as excellent electrical conductivity, electrochemically active surface area, thermal conductivity, good mechanical strength, and optical transmittance [15–20]. Graphene can be combined with different metals to make its composites for supercapacitor electrodes. In this chapter, we mainly focus on different graphene-based composite electrode materials for supercapacitor applications.

2. Graphene based nanocomposites for supercapacitors electrodes

Graphene and its composites are the most attractive choices for increasing the functionality of supercapacitors through improving electrode conduction characteristics. Graphene has been evaluated for high-performance supercapacitor electrodes, whether as a single layer or as a beneficial component to make its nanocomposite for electrodes [21]. Composite materials are materials that combine two or even more materials with different characteristics to generate a final product consisting of remarkable properties. As previously stated, graphene has a variety of different attributions, any number of which might be exploited to create excellent composites. The presence of graphene allows for the creation of composites with exceptional properties, which ultimately improves the conductivity and mechanical strength of bulk material [22]. To make high-quality composites, graphene may be combined with polymers metals, ceramics, and some other active materials. Graphene composites have such a broad range of applications, and much research is being conducted to develop novel and fascinating materials.

2.1 Graphene/manganese dioxide composite

MnO_2 is regarded as a significant electrode option because of its significant specific capacity of about 1370 F g^{-1}, relatively inexpensive, and nontoxicity [4, 23–25]. Unfortunately, the conductivity of MnO_2 is low, and the practical specific capacitance is substantially lower than the theoretical value-specific capacitance [26]. Graphene a 2D material has the ability to enhance the electrical conductivity of MnO_2-based electrodes [27]. The large surface area and outstanding electrical conductivity of graphene enhance the interconnectivity of conductive paths for MnO_2, and the subsequent MnO_2/graphene nanocomposites demonstrated better electrochemical capacitive performance and increases the performance of energy production and storage devices.

Xia Li et al. used the electrostatic self-assembly technique to synthesize HG/MnO_2 composites. A holey approach for graphene opens up additional ion channels, improving rate capability in supercapacitors. Furthermore, the electrostatic self-assembly of holey graphene and MnO_2 nanosheets significantly improves electron transport channels, leading to a high value of specific capacitance. Furthermore, the electrostatic self-assembly technique has the ability to regulate the mass ratio of graphene and MnO_2 throughout the final composite. At 0.5 A/g, the obtained HG/MnO_2 composite electrode containing a 14.8% holey graphene composition shows a specific capacitance of about 219.3 F g^{-1}, and also at 10 A/g, it shows good capacity retention of more than 61.4%. [28].

M. Zhang et al. [29] Developed a simple alcohol-improvement methodology based on electrodeposition to synthesize MnO_2 and MnO_2/RGO composite electrode materials. Following alcohol treatment, the accumulated MnO_2 on the substrate has even more homogeneous thickness as well as looser dispersion, whereas RGO has a broader depositing area having a higher compact distribution. Furthermore, the rate of capability, cycle performance, specific energy, and specific power have all been significantly enhanced. At 1 A/g, the prepared MnO_2 electrode shows a large value of specific capacitance about 270 F/g having a retention of a capacity was approximately 83.9% and also the MnO_2/RGO electrode exhibits a good specific capacity of about 467 F/g at 1/g having 93.1% capacity retention was achieved after 2500 cycles.

M. Zhang et al [30]. used a solution-based ultrasonic-assisted technique to prepare graphene-MnO2 composite. The porous MnO_2 microspheres coated in graphene nanosheets with such a large specific surface area, allowing for rapid ion diffusion/transport. At 0.5 A/g the graphene-MnO_2 composite electrode shows an excellent electrochemical performance having a specific capacity of about 1227 F/g and keeping capacity retention at approximately 90% of its first cycle. Furthermore, the asymmetric supercapacitor depending mostly on graphene-MnO_2 composite exhibits a large energy density of about 19.6 Wh kg^{-1} at even a power density of about 351 W kg^{-1}, indicating a strong prospective as an electrode material supercapacitor.

2.2 Iron oxide-graphene composite

Iron oxides (FeOx) are attractive electrodes among transition metal oxides due to their large capacitance, low price, varied oxidation states, and environment friendly nature [31–34]. While the poor conductivity of iron oxide has hindered its super capacitive applicability, combining Fe_3O_4 with graphene can be regarded as a viable solution technique. [35]. Graphene has the ability to improve the performance of iron oxide-based electrodes.

Siyu Su et al. [36] prepared Fe_3O_4/NG composite by using a simple one-step green and scalable dry technique. The generation of C-O-Fe bond shows that Fe_3O_4 nanoparticles are closely linked with the layers of graphene, resulting in good stability and electrical conductivity. The obtained Fe_3O_4/NG composite exhibits remarkable electrochemical capabilities, including a large value of specific capacitance of about 740 F/g even at 1 A/g, exceptional cycle stability, and significant rate capability.

A.J. Khan et al. [37] effectively manufactured nanodisc-shaped Fe_3O_4/rGO composites using a quick and simple hydrothermal technique and a brief annealing procedure. The electrochemical characteristics of a Fe_3O_4/rGO composite have been improved for use as electrode material for supercapacitor applications. The prepared Fe_3O_4/rGO composite outperforms the Fe_3O_4 nanodiscs in terms of electrochemical operation, achieving a high specific capacitance of about 1149 F/g at 1.5 A/g. Furthermore, after conducting consecutive 10,000 cycles at 10 A/g, the obtained Fe3O4/rGO composite demonstrates good cyclic stability of 97.53% and amazing rate capability of 87%. **Figure 1** shows the FESEM and TEM images of the Fe_3O_4/rGO composite.

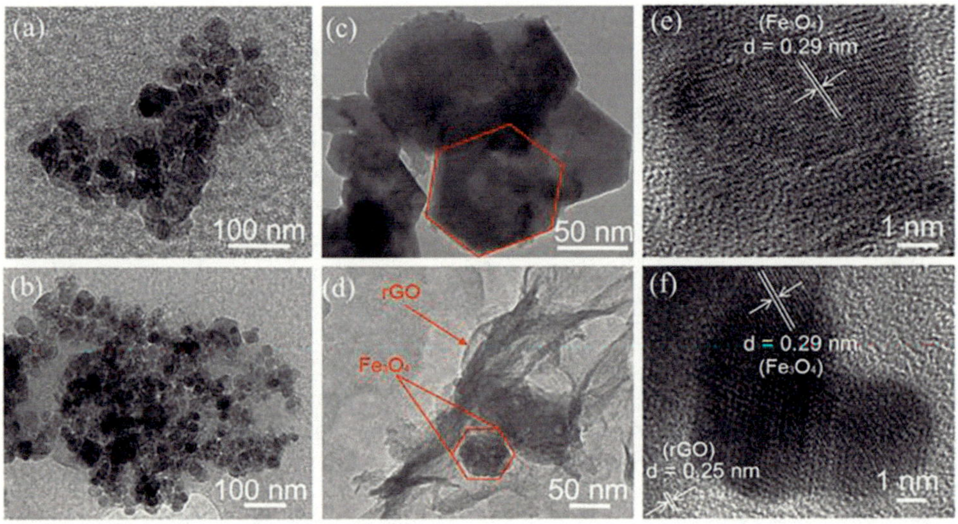

Figure 1.
(a, b) FESEM pictures of Fe_3O_4/rGO nanodiscs electrode composite at various resolutions; (c, d) TEM picture of Fe_3O_4/rGO composite; and (e, f) HRTEM pictures of Fe3O_4 nanodiscs and Fe_3O_4/rGO nanodiscs. Reprinted with permission from ref [37] Elsevier Copyright @ 2020.

Siyu Su et al. [38] used an effective and sustainable pyrolysis knowledge to produce nitrogen-loaded porous graphene decorated with Fe_3O_4 nanoparticles (NPGF). The NPGF nanohybrids have an equally dispersed pore architecture and a very pure content that is free of contaminants, according to structural and compositional evaluation. Furthermore, electrochemical characterization confirms the excellent electrochemical efficiency, which includes specific capacitance of about 713 F g^{-1} even at 1 A g^{-1}, remarkable capability rate having retention of capacity of about 77.3% and also shows best capacity retention of approximately 67.9% whenever the current density is rapidly increased from 1–10 and 20 A g^{-1}, as well as exceptional cycling stability (after 3000 it shows a capacitance retention about 94.3% .

By using a chemical reduction-high-temperature process, Zhang et al. [39] created graphene/Fe3O_4 (GN/Fe_3O_4) nanocomposites as the electrode of supercapacitors. The surface of graphene is routinely decorated with Fe_3O_4 particles of equal size. At 0.5 A/g, the obtained GN/Fe_3O_4 composite electrode delivers a good specific capacitance about 265.6 F/g. Ultimately, the button supercapacitors were built using the produced composite materials as electrodes. After 100 consecutive charging / discharging cycles, the nanocomposites demonstrate constant capacitance efficiency. After 500 charging/discharging cycles, the capacitance efficiency remains above 80%, suggesting that the nanocomposite has outstanding cycle stability.

Mustafa Aghazadeh et al. [40] used a simple one-step electrochemical approach to create 3D N-combined porous graphene/magnetite nanoparticles hybrids on nickel foam (Fe_3O_4/3D-NPG/NF electrode). Three-dimensional nitrogen combined with porous graphene are electrophoretically synthesized on Ni foam in this process, while magnetite particles are electrochemically deposited on the surface of 3D-NPG layers. In contrast, clean Fe_3O_4/NF and 3D-NPG/NF composite electrodes was produced by depositing Fe_3O_4 particles and N-doped graphene individually over Ni foam. At 2 A/g, after 5000 GCD cycles, the obtained Fe_3O_4/3D-NPG composite electrode shows good specific capacity value about 715 F/g and also delivers cycle life about 94.3%.

2.3 Cobalt oxide—graphene composites

Co_3O_4 is one of the extensively investigated transition metal oxides for supercapacitor applications because of its long-term cycle performance, wide surface area, high conductivity, superior corrosion resistance, and natural availability [41–45]. However, Co_3O_4 has weaker electronic conductivity dramatically reduces its practical applicability. To address this issue, through a growth mechanism, Co_3O_4 is mixed with electrically conductive carbon-based materials to produce a hybrid nanocomposite [46, 47]. Because of its large surface area and exceptional electrical conductivity, graphene is the most suitable candidate to combine with Co_3O_4 to enhance the electrochemical characteristics of the Co_3O_4 electrode.

Venkatachalam et al. [48] used a hydrothermal technique to produce a 1D/2D Co_3O_4/rGO composite. The Co_3O_4/rGO composite electrode demonstrated greater specific capacitance and improved cycle stability than the conventional Co_3O_4 electrode. At 0.5 A g^{-1}, the composite material provided an exceptional supercapacitance of about 916.6 F g^{-1}.

S. Sagadevan et al. [49] used a straightforward hydrothermal method to synthesize a Co_3O_4/rGO nanocomposite that was then used as an appropriate electrode in supercapacitors. The addition of rGO in Co_3O_4 improved the electrochemical performance of the produced Co_3O_4/rGO nanocomposite. The Co_3O_4/rGO composite-based supercapacitor obtained a maximum specific capacitance of about 754 Fg1 as well as outstanding stability, holding 96% capacity after 1000 consecutive cycles. The reduction in surface area of graphene sheets is responsible for the stability, rise in specific capacity, wettability, and stability of electrode materials. As a result, the electrochemical behavior of Co_3O_4/rGO nanocomposite makes it an obvious option for a high-performance supercapacitor. **Figure 2** shows SEM pictures of graphene oxide, Co_3O_4, and Co_3O_4/rGO composite.

M. Yun et al. [50] used a one-step hydrothermal assembly procedure to make a high-capacitive Co_3O_4/HG composite electrode. Holey graphene offers a suitable framework for high electrical conductivity in the prepared electrode; the numerous holes in holey graphene give a shorter ion channel, and the Co_3O_4 nanoparticles encapsulated in the holey graphene network give high capacitance. As a consequence, the Co_3O_4/HG composite electrode shows excellent performance as compared to the pure Co_3O_4 electrode. At 1 A/g, the Co_3O_4/HG composite-based supercapacitor provided a high value of specific capacity of about 825 F g^{-1} at 1 A g^{-1} with higher rate capacity after further refining mass ratio of Co_3O_4 to holey graphene.

Y. Jiang et al. [51] used a simple and cost-effective method for the preparation of graphene/Co_3O_4 composite through one-pot ball-milling of graphite, $Co(CH_3COO)_2$ and $(NH_4)_2CO_3$ (**Figure 3**). Graphite was inserted by $(NH_4)_2CO_3$ and simultaneously separated into a few layers of graphene sheets having consisted of functional groups during the process. Mechanochemical interactions involving $(NH_4)_2CO_3$ and $Co(CH_3COO)_2$ were then used to generate Co_3O_4 nanocrystals on the interface of graphene sheets. The two components' synergistic effects increase electron/ion conduction and capacitive responses, resulting in outstanding rate performance and a large value of specific capacitance for graphene/Co_3O_4 composite electrodes.

2.4 Nickel oxide- graphene composite

Nickel oxide (NiO) is considered to be one of the best electrode materials because of its affordable cost, non-toxicity, good chemical/thermal stability, easily accessible, and eco-friendly nature [52–58]. Oxides of Ni are highly fascinating because

Figure 2.
SEM pictures of (a) Graphene oxide; (b) Co_3O_4; and (c & d) Co_3O_4/rGO nanocomposite. Reprinted with permission from ref [49] Elsevier Copyright @2020.

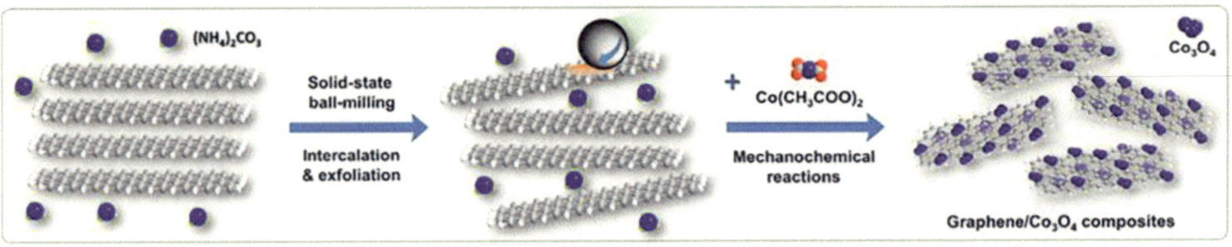

Figure 3.
Synthesis procedure of graphene/Co_3O_4 composites electrode by ball-milling technique. Reprinted with permission from ref [51] Elsevier Copyright @2020.

of numerous oxidation states exhibited by Ni [59]. Combining two-dimensional NiO nanosheets with graphene can increase electrical conductivity of the material, which has proven to be a very successful method of improving the electrochemical characteristics of NiO-based electrode materials [60].

Y. Zhang et al. [61] used an electrode-assisted plasma electrolysis technique to produce rGO/NiO composite electrodes. This approach is simple and quick, allowing for one-step production of rGO/NiO composites. At 1 A/g^{-1}, the obtained rGO/NiO composite exhibits a good value of specific capacitance of about 1093 F g^{-1}. It also has good cycle stability, as well as coulombic efficiency about 90.6% was retained after 5000 cycles. **Figure 4** shows the preparation procedure for rGO/NiO composite and HRTEM images of rGO/NiO composite.

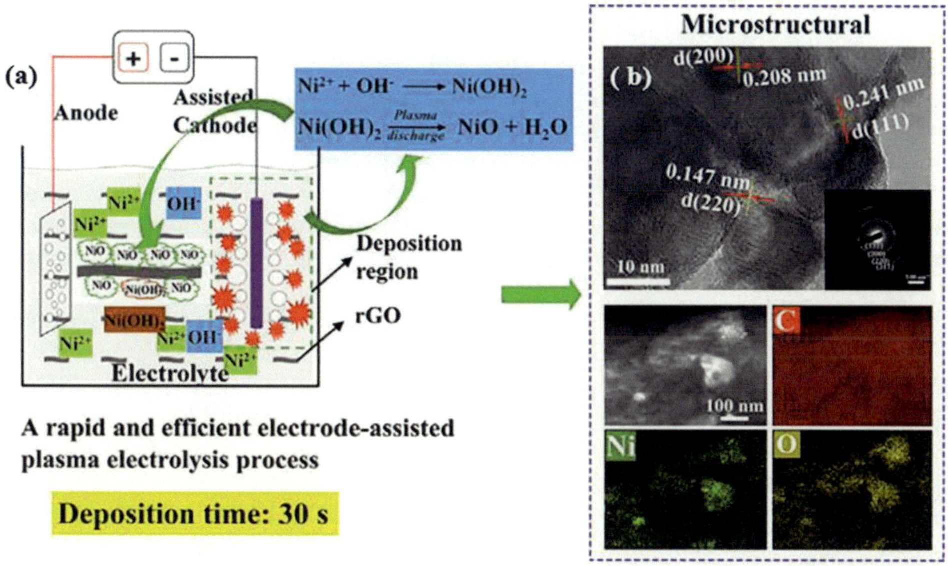

Figure 4.
(a) The electrode-assisted plasma electrolysis method equipment; (b) rGO/NiO composite electrode HRTEM image; and (c) The SAED profile of the prepared rGO/NiO composite electrode is shown in the inset. Images of rGO/NiO composite. Different EDS-elemental mapping: of C, Ni, and O elements. Reprinted with permission from ref [61] Elsevier Copyright @ 2020.

M. Sethi et al. [62] developed a solvothermal technique to easily synthesize porous graphene-NiO (PGNO) electrodes using a mixed solvent solution. When the PG is properly loaded onto the NiO nanoflakes, a stronger composite structure is formed, which contributes to resisting volumetric changes throughout electrochemical cycling. The obtained composite material shows an amazing specific capacitance of about 511.0 F g^{-1} for the one electrode at 5 mV s^{-1}, good power density, and excellent cycle stability. The excellent electrochemical performance of this composite is principally attributed to the synergistic effect of the constituent materials, which offered an effective ion reservoir as well as mechanical strength, allowing for continuous transfer of electrolyte ions during much of the electroactive material once confined to charge and discharge processes.

Shu-xia Yuan et al. [63] used heterogeneous self-assembly to create a GO/$Ni(HCO_3)_2$ composite, which was then thermally treated to produce rGO/NiO. The prepared rGO/NiO composite has a porous volume of about 0.26 cm^3 g^{-1} and a larger specific surface area (121.3 m^2 $g^{-1)}$ and porous volume of about 0.26 cm^3 g^{-1}. Because of the release of H_2O and CO_2, The hierarchical porosity dispersion of rGO/NiO composite is around 2–100 nm. Because of its porosity dispersion and large specific surface area, the rGO/NiO composite shows a larger value of specific capacitance of about 919 F g^{-1} at 0.5 A g^{-1}, and an enhanced rate capability of about 71% was achieved whenever the current density is increased from 0.5 to 5 A g^{-1}.

2.5 Zinc oxide- graphene composite

ZnO is a highly interesting electrode material due to its affordable cost, environmentally friendly nature, nontoxicity, availability, high specific energy, and excellent electrochemical stability [64–68]. Because of the synergistic interaction between the constituents, a novel nanocomposite made of ZnO, and graphene is predicted to have unique characteristics and capabilities. Many innovative techniques have been developed to produce graphene-based zinc oxide nanocomposites and their potentially advantageous properties [67]. Graphene-based ZnO composites electrode has the ability to improve the performance of supercapacitor. R Kumar et al. [69] used a simple and rapid microwave approach to synthesize ZnO NPs and bind them to rGO-NSs surfaces to produce ZnO/rGO nanocomposite for supercapacitor electrode application. At 30 mV/s, the electrode of ZnO/rGO nanocomposite displayed a good specific capacitance of about 102.4 F g^{-1}. After 3000 cycles, the capacitance stability was reattained at 82.5%.

M miah et al. [70] used a simple ex-situ wet chemical procedure to produce porous ZnO nanospheres implanted on rGO (ZnO/rGO composite). At 1 A g^{-1}, the ZnO/rGO composite electrodes deliver a good specific capacitance of about 949 F g^{-1}. The ZnO/rGO composite also has remarkable cyclic stability, retaining 91% of its original capacitance during 10000 cycles. The pseudocapacitive effect of very porous ZnO nanospheres and the EDLCs of reduced graphene oxide increase specific capacitance.

Jianping Xu et al. [71] used a straightforward hydrothermal method to produce ZnO nanorods/NG composite electrodes. ZnO nanorods with length of 1–2 m and diameters 50–100nm was supported uniformly distributed on NG networks. The prepared ZnO/NG composite has good electrochemical performance, having a large value of specific capacity of about 237.18 F/g at 0.5 A/g, rate capability of about 150 F/g at 10 A/g, and after 2000 cycles it shows long-term stability of about 90.8%.

3. Optimization of graphene electrode-electrolyte for supercapacitors

Carbon materials with large surface areas for charge storage are now the most researched materials. Despite these huge precise surface areas, the charges substantially deposited on the carbon electrodes. The deposition is somehow restricted by many factors, such as volume expansion, layer formation on the surface, surface-to-volume ratio, etc. [72]. Therefore, an alternate electrode material is needed to enhance the performance of supercapacitor electrodes. Metal oxides and graphene is considered to be the exceptional electrode ingredients to enhance the performance of supercapacitors.

Electrolyte parameters, particularly specific capacitance and energy density are critical in determining the capacitive efficiency of electric double-layer capacitors (EDLCs). In particular, electrolytes with such a large electrochemical stability window (ESW) can provide superior specific capacitance as well as density, and that is why ionic liquid-based electrolytes have received a lot of attention. The quantity of ionic liquid (IL) is a significant parameter for controlling its viscosity, potential window, and ionic conductivity, which is represented in the EDLC working voltage and has yet to be well investigated [73]. A pseudocapacitor achieves charge aggregation by faradic reactions (redox reactions) of redox-active materials deposited on the electrodes or immersed in fluids. Because of their excellent redox characteristics, metal oxides have frequently been employed as cathodes for those deposited on the electrodes [74].

Graphene-based micro-supercapacitors on standard Xerox paper substrates were created in response to future demands for flexible, simple, and moderate energy storage devices. To enhance the overall performance of the device, the usage of redox-active species (iodine redox couple) was investigated. At 6.5 mA cm^{-3}, the smart printed device based on graphene composite had an amazing high volumetric capacitance of 29.6 mF cm^{-3} (volume of entire device). These electrodes contain redox-active potassium iodide. Surprisingly, the device demonstrated enhanced volumetric capacitance of 130 mF cm^{-3}. In an H_2SO_4 solution, the maximal density for a graphene +K device was determined to be 0.026 mWh cm^{-3} [75].

Surjit Sahoo et al. [76] described a unique supercapacitor device that uses 2D graphene sheets as electrode material. The markedly high energy storage and a porosity PVDF electrolyte incorporating $TEABF_4$ as a solid-like piezo-polymer separator were used. When placed under stress, stresses ranging from 5 to 20 N, the porous PVDF film produced a voltage ranging from 4 to 11 V. The graphene polymer composite device had the highest particular device capacitance of 28.46 F g^{-1} (31.63 mF cm^{-2}) as well a specific energy of 35.58 Wh kg^{-1}, as well as a great power density of 7500 W kg^{-1}. The above discussion shows that the optimum composition of electrolyte, as well as electrode configuration, is very important to obtain high capacity as well as cyclability.

4. Conclusion

Regarding graphene's amazing and unique properties, it is one of the best-known materials that may be applied for a variety of purposes. Graphene can be applied as electrode material in energy storage devices including rechargeable batteries and supercapacitors to store, absorb, and release large amounts of energy. Moreover, graphene-based metal oxides have the ability to increase the energy density, capacity, and performance of supercapacitors. Metal oxides have been considered one of the best electrode materials for supercapacitor application due to their affordable cost, easy synthesis, high capacity, environment friendly nature, and non-toxicity. However, conventional metal oxides have many disadvantages which limit their practical application including weak cyclic stability, low electrical conductivity, low power, and energy density. Graphene a 2D has the ability to solve these problems. Graphene can be combined with these metal oxides to make its composites to improve the performance of supercapacitors. Moreover, much attention has been focused in recent years on structural architecture, material production, and device performance assessment. To achieve the predicted full-scale actual use, the electrode materials' efficiency, as well as reproducible quantity, must be increased in the near future.

Author details

Kefayat Ullah[1]*, Bakht Mand Khan[1], Amin Ur Rashid[1] and Won Chun Oh[2]

1 Department of Applied Physical and Material Sciences, University of Swat, Swat, Pakistan

2 Department of Advanced Materials Science and Engineering, Chungnam, South Korea

*Address all correspondence to: kefsayed@uswat.edu.pk

IntechOpen

© 2022 The Author(s). Licensee IntechOpen. This chapter is distributed under the terms of the Creative Commons Attribution License (http://creativecommons.org/licenses/by/3.0), which permits unrestricted use, distribution, and reproduction in any medium, provided the original work is properly cited.

References

[1] Mane V et al. Enhanced specific energy of silver-doped MnO_2/graphene oxide electrodes as facile fabrication symmetric supercapacitor device. Materials Today Chemistry. 2021;**20**:100473

[2] Park J et al. Graphene-based two-dimensional mesoporous materials: Synthesis and electrochemical energy storage applications. Materials. 2021;**14**(10):2597

[3] Paraschiv C et al. Hydrothermal growth of ZnO/GO hybrid as an efficient electrode material for supercapacitor applications. Scripta Materialia. 2021;**195**:113708

[4] Zhu C et al. Direct laser writing of MnO_2 decorated graphene as flexible supercapacitor electrodes. Journal of Materials Science. 2020;**55**(36):17108-17119

[5] Yao J et al. Ternary flower-sphere-like MnO_2-graphite/reduced graphene oxide nanocomposites for supercapacitor. Nanotechnology. 2021;**32**(18):185401

[6] Lee SJ et al. Heteroatom-doped graphene-based materials for sustainable energy applications: A review. Renewable and Sustainable Energy Reviews. 2021;**143**:110849

[7] Velasco A et al. Recent trends in graphene supercapacitors: From large area to microsupercapacitors. Sustainable Energy and Fuels. 2021;**5**(5):1235-1254

[8] Wang Y et al. Flexible supercapacitor: Overview and outlooks. Journal of Energy Storage. 2021;**42**:103053

[9] Xiong S et al. A high-performance hybrid supercapacitor with NiO derived NiO@ Ni-MOF composite electrodes. Electrochimica Acta. 2020;**340**:135956

[10] Mohanty A et al. An extensive review on three dimension architectural metal-organic frameworks towards supercapacitor application. Journal of Power Sources. 2021;**488**:229444

[11] Adib K et al. Sonochemical synthesis of Ag_2WO_4/RGO-based nanocomposite as a potential material for supercapacitors electrodes. Ceramics International. 2021;**47**(10):14075-14086

[12] Tale B, Nemade K, Tekade P. Graphene based nano-composites for efficient energy conversion and storage in solar cells and supercapacitors: A review. Polymer-Plastics Technology and Materials. 2021;**60**(7):784-797

[13] Arvas MB, Gencten M, Sahin Y. One-step synthesized N-doped graphene-based electrode materials for supercapacitor applications. Ionics. 2021;**27**(5):2241-2256

[14] Ma Y et al. Recent advances in transition metal oxides with different dimensions as electrodes for high-performance supercapacitors. Advanced Composites and Hybrid Materials. 2021;**4**(4):906-924

[15] Zheng X et al. Graphene-based fibers for the energy devices application: A comprehensive review. Materials & Design. 2021;**201**:109476

[16] Wu X, Mu F, Lin Z. Three-dimensional printing of graphene-based materials and the application in energy storage. Materials Today Advances. 2021;**11**:100157

[17] Zhang B et al. Computational screening toward quantum capacitance

of transition-metals and vacancy doped/co-doped graphene as electrode of supercapacitors. Electrochimica Acta. 2021;**385**:138432

[18] Iqbal AA et al. Graphene-based nanocomposites and their fabrication, mechanical properties and applications. Materialia. 2020;**12**:100815

[19] Bai Y, Xu T, Zhang X. Graphene-based biosensors for detection of biomarkers. Micromachines. 2020;**11**(1):60

[20] Ullah K et al. Electrochemical performance of graphene/activated carbon based electric double layer supercapacitors. Asian Journal of Chemistry. 2016;**28**(1):133

[21] Tiwari SK et al. Current research of graphene-based nanocomposites and their application for supercapacitors. Nanomaterials. 2020;**10**(10):2046

[22] Hussain SZ et al. A review on graphene based transition metal oxide composites and its application towards supercapacitor electrodes. SN Applied Sciences. 2020;**2**(4):1-23

[23] Zhou Y et al. Hierarchically structured electrodes for moldable supercapacitors by synergistically hybridizing vertical graphene nanosheets and MnO_2. Carbon. 2021;**172**:272-282

[24] Wu D et al. MnO_2/carbon composites for supercapacitor: Synthesis and electrochemical performance. Frontiers in Materials. 2020;**7**:2

[25] Bai X-L et al. Supercapacitor performance of 3D-graphene/MnO_2 foam synthesized via the combination of chemical vapor deposition with hydrothermal method. Applied Physics Letters. 2020;**117**(18):183901

[26] Zhang M, Yang D, Li J. Supercapacitor performances of MnO_2 and MnO_2/reduced graphene oxide prepared with various electrodeposition time. Vacuum. 2020;**178**:109455

[27] Chang H-W et al. Soft X-ray absorption spectroscopic investigation of MnO_2/graphene nanocomposites used in supercapacitor. Catalysis Today. 2022;**388**:63-69

[28] Li X et al. Holey graphene/MnO_2 nanosheets with open ion channels for high-performance solid-state asymmetric supercapacitors. International Journal of Energy Research. 2020;**44**(5):3446-3457

[29] Zhang M, Yang D, Li J. Effective improvement of electrochemical performance of electrodeposited MnO_2 and MnO_2/reduced graphene oxide supercapacitor materials by alcohol pretreatment. Journal of Energy Storage. 2020;**30**:101511

[30] Zhang M et al. Graphene-wrapped MnO_2 achieved by ultrasonic-assisted synthesis applicable for hybrid high-energy supercapacitors. Vacuum. 2020;**176**:109315

[31] Gaire M et al. Flexible iron oxide supercapacitor electrodes by photonic processing. Journal of Materials Research. 2021;**36**(22):4536-4546

[32] Cai D et al. Iron oxide nanoneedles anchored on N-doped carbon nanoarrays as an electrode for high-performance hybrid supercapacitor. ACS Applied Energy Materials. 2020;**3**(12):12162-12171

[33] Wang Y et al. Single-step preparation of ultrasmall iron oxide-embedded carbon nanotubes on carbon cloth with excellent superhydrophilicity and enhanced supercapacitor performance. ACS Applied Materials and Interfaces. 2021;**13**(38):45670-45678

[34] Ms NAD et al. Investigation of chemical bonding and supercapacitivity properties of Fe_3O_4-rGO nanocomposites for supercapacitor applications. Diamond and Related Materials. 2020;**104**:107756

[35] Barmi A-AM et al. Binder-free high-performance Fe_3O_4 fine particles in situ grown onto N-doped porous graphene layers co-embedded into porous substrate as supercapacitor electrode. Journal of Materials Science: Materials in Electronics. 2020;**31**(18):15198-15217

[36] Su S et al. One-step green and scalable dry synthesis of nitrogen-doped graphene-encapsulated Fe_3O_4 nanoparticles as high-performance supercapacitor electrode. Journal of Alloys and Compounds. 2020;**834**:154477

[37] Khan AJ et al. Surface assembly of Fe_3O_4 nanodiscs embedded in reduced graphene oxide as a high-performance negative electrode for supercapacitors. Ceramics International. 2020;**46**(11):19499-19505

[38] Su S et al. Nitrogen-doped porous graphene coated with Fe_3O_4 nanoparticles for advanced supercapacitor electrode material with improved electrochemical performance. Particle and Particle Systems Characterization. 2020;**37**(4):2000011

[39] Zhang J et al. The graphene/Fe_3O_4 nanocomposites as electrode materials of supercapacitors. Journal of Nanoscience and Nanotechnology. 2020;**20**(5):3164-3173

[40] Aghazadeh M et al. On-pot fabrication of binder-free composite of iron oxide grown onto porous N-doped graphene layers with outstanding charge storage performance for supercapacitors. Journal of Materials Science: Materials in Electronics. 2021;**32**(10):13156-13176

[41] Yadav S, Devi A. Recent advancements of metal oxides/nitrogen-doped graphene nanocomposites for supercapacitor electrode materials. Journal of Energy Storage. 2020;**30**:101486

[42] Lakra R et al. Facile synthesis of cobalt oxide and graphene nanosheets nanocomposite for aqueous supercapacitor application. Carbon Trends. 2022;**7**:100144

[43] Yetiman S et al. Microwave-assisted fabrication of high-performance supercapacitors based on electrodes composed of cobalt oxide decorated with reduced graphene oxide and carbon dots. Journal of Energy Storage. 2022;**49**:104103

[44] Zhu M et al. Glycerol-assisted tuning of the phase and morphology of iron oxide nanostructures for supercapacitor electrode materials. Materials Chemistry Frontiers. 2021;**5**(6):2758-2770

[45] Zhu X et al. Simply synthesized N-doped carbon supporting Fe_3O_4 nanocomposite for high performance supercapacitor. Journal of Alloys and Compounds. 2020;**821**:153580

[46] Jadhav S et al. Probing electrochemical charge storage of 3D porous hierarchical cobalt oxide decorated rGO in ultra-high-performance supercapacitor. Surface and Coatings Technology. 2021;**419**:127287

[47] Li S et al. Three-dimensional porous carbon/Co_3O_4 composites derived from graphene/Co-MOF for high performance supercapacitor electrodes. Applied Surface Science. 2020;**503**:144090

[48] Venkatachalam V, Jayavel R. 1D/2D Co_3O_4/graphene composite electrodes for high-performance supercapacitor applications. Journal of Electronic Materials. 2020;**49**(5):3174-3181

[49] Sagadevan S et al. Reduced graphene/nanostructured cobalt oxide nanocomposite for enhanced electrochemical performance of supercapacitor applications. Journal of Colloid and Interface Science. 2020;**558**:68-77

[50] Yun M et al. Holey graphene interpenetrating networks for boosting high-capacitive Co_3O_4 electrodes via an electrophoretic deposition process. Ceramics International. 2021;**47**(19):27210-27216

[51] Jiang Y et al. Scalable mechanochemical coupling of homogeneous Co_3O_4 nanocrystals onto in-situ exfoliated graphene sheets for asymmetric supercapacitors. Chemical Engineering Journal. 2020;**397**:125503

[52] Dhas SD et al. Synthesis of NiO nanoparticles for supercapacitor application as an efficient electrode material. Vacuum. 2020;**181**:109646

[53] Zhou S et al. An electrochromic supercapacitor based on an MOF derived hierarchical-porous NiO film. Nanoscale. 2020;**12**(16):8934-8941

[54] Wang J et al. Construction of hierarchical Co9S8@ NiO synergistic microstructure for high-performance asymmetric supercapacitor. Journal of Colloid and Interface Science. 2021;**603**:440-449

[55] Ahmed R, Nabi G. Enhanced electrochemical performance of Cr-doped NiO nanorods for supercapacitor application. Journal of Energy Storage. 2021;**33**:102115

[56] Sannasi V et al. H_2O_2-assisted microwave synthesis of NiO/CNT nanocomposite material for supercapacitor applications. Ionics. 2020;**26**(8):4067-4079

[57] Obodo RM et al. Conjugated NiO-ZnO/GO nanocomposite powder for applications in supercapacitor electrodes material. International Journal of Energy Research. 2020;**44**(4):3192-3202

[58] Shi H et al. Preparation of petal-particle cross-linking flowerlike NiO for supercapacitor application. Journal of Electroanalytical Chemistry. 2020;**876**:114481

[59] Sethi M, Shenoy US, Bhat DK. Hassle-free solvothermal synthesis of NiO nanoflakes for supercapacitor application. Physica B: Condensed Matter. 2021;**611**:412959

[60] Gao X et al. Hybrid two-dimensional nickel oxide-reduced graphene oxide nanosheets for supercapacitor electrodes. Microchemical Journal. 2021;**164**:105979

[61] Zhang Y et al. One-step synthesis of the reduced graphene oxide@ NiO composites for supercapacitor electrodes by electrode-assisted plasma electrolysis. Materials & Design. 2020;**196**:109111

[62] Sethi M, Shenoy US, Bhat DK. Simple solvothermal synthesis of porous graphene-NiO nanocomposites with high cyclic stability for supercapacitor application. Journal of Alloys and Compounds. 2021;**854**:157190

[63] Yuan S-X et al. Synthesis of a rGO/NiO composite with a hierarchical porous structure by self-assemblyand its electrochemical performance as a supercapacitor electrode. New Carbon Materials. 2020;**35**(6):731-738

[64] Rabani I et al. The role of uniformly distributed ZnO nanoparticles on cellulose nanofibers in flexible solid state symmetric supercapacitors. Journal of Materials Chemistry A. 2021;**9**(19):11580-11594

[65] Angelin MD et al. Electrochemical investigation of Zr-doped ZnO nanostructured electrode material for high-performance supercapacitor. Ionics. 2020;**26**(11):5757-5772

[66] Chen X et al. A novel strategy of multi-element nanocomposite synthesis for high performance ZnO-$CoSe_2$ supercapacitor material development. Chinese Journal of Chemistry. 2021;**39**(9):2441-2450

[67] Bhat U, Meti S. Graphene-based ZnO nanocomposites for supercapacitor applications. Materials Research Foundations. 2020;**64**:181

[68] Murali S et al. Polyol mediated synthesis of anisotropic ZnO nanomaterials and composite with rGO: Application towards hybrid supercapacitor. Journal of Alloys and Compounds. 2020;**844**:156149

[69] Kumar R et al. One-pot synthesis of reduced graphene oxide nanosheets anchored ZnO nanoparticles via microwave approach for electrochemical performance as supercapacitor electrode. Journal of Materials Science: Materials in Electronics. 2020;**31**(18):15456-15465

[70] Miah M et al. Study of highly porous ZnO nanospheres embedded reduced graphene oxide for high performance supercapacitor application. Electrochimica Acta. 2020;**354**:136675

[71] Xu J et al. A facile synthesis of ZnO nanorods on nitrogen-doped graphene sheets for supercapacitor applications. International Journal of Electrochemical Science. 2020;**15**:765-773

[72] Qi Z et al. Nanostructured metal oxides-based electrode in supercapacitor applications. Supercapacitor Design and Applications. 2016;**7**:411-456

[73] Wong SI et al. Optimization of ionic-liquid based electrolyte concentration for high-energy density graphene supercapacitors. Applied Materials Today. 2020;**18**:100522

[74] Guo C et al. High-performance asymmetric supercapacitors using holey graphene electrodes and redox electrolytes. Carbon. 2020;**157**:298-307

[75] Nagar B et al. Design and fabrication of printed paper-based hybrid micro-supercapacitor by using graphene and redox-active electrolyte. ChemSusChem. 2018;**11**(11):1849-1856

[76] Sahoo S et al. High performance self-charging supercapacitors using a porous PVDF-ionic liquid electrolyte sandwiched between two-dimensional graphene electrodes. Journal of Materials Chemistry A. 2019;**7**(38):21693-21703

Chapter 4

Review on Transition Metal Oxides and Their Composites for Energy Storage Application

Nithya S. George, Lolly Maria Jose and Arun Aravind

Abstract

Supercapacitors evolved as a breakthrough to the existing shortages in energy resources because of its enhanced capacitive performance, long-term stability, and high power density. Transition metal oxides (TMOs), a redox active material in energy storage applications, showing high specific capacitance (100–2000 F/g) than the electrical double-layer capacitor (EDLC) material has been reviewed a lot. Among various TMOs, nickel oxide (NiO), tin oxide (SnO2), manganese dioxide (MnO2), tungsten oxide (WO3), vanadium pentoxide (V2O5) are widely used by researchers due to their high theoretical capacitance, low cost, and long cycle life. The limitations of TMO-based electrode material includes low electrical conductivity, ion mobility, and low energy density. It is thus important to develop proper combination of TMO with other transition metals, TMOs, transition metal dichalcogenides (TMDs), conducting polymers (CPs) and carbon-based materials (graphene oxide (GO), activated carbon (AC) and reduced GO (rGO)). This chapter focuses on ongoing development in six TMO-based electrode material (NiO, ZnO, MnO2, SnO2, WO3, V2O5) fabrication for the enhancement of electrochemical performance, their synthesis method and then review about the recent progress in studying the supercapacitor performance of the material. The limitations of each TMOs listed separately, providing new insights for future energy storage applications.

Keywords: transition metal oxides, nanocomposites, transition metal dichalcogenides, supercapacitor

1. Introduction

It is a matter of great concern that the conventional energy sources on our planet are getting exhausted day by day. As the energy consumption demand is constantly increasing, new alternative energy sources are being developed across the world [1–5]. The highly threatening increase in pollution level and global warming invites our attention to the necessity of developing a clean energy portfolio. Special focus should be given in the development of alternative energy sources as well as its storage [6]. The most well-known energy production and storage technologies are batteries, fuel cells, and supercapacitors. In contrast to fuel cells and batteries, supercapacitors use a different energy generating technique. Even though these three systems have distinct energy storage and conversion processes, there exists some electrochemical similarities between them. Common characteristics include the separation of electron and ion transport and the fact that the energy-producing activities occur at the electrode/electrolyte interface's phase boundary. Also, the basic structure of these three systems consists of two electrodes in contact with an electrolyte solution [7].

Coming to the differences, in fuel cells and batteries, energy production occurs from chemical reaction via redox reaction, whereas in supercapacitors, energy is liberated through the diffusion of ions at the interface between the electrolyte and the inner side of the capacitor electrode plates forming electrostatic double layers. Supercapacitors are also called ultracapacitors. They act as a bridge between conventional capacitors and batteries. Conventional capacitors deliver high power density with low energy density and the batteries possess high energy density with low power density. Supercapacitors are generally preferred over batteries because they enable quick charging and can deliver energy at a pace that is comparably faster. Also, they are acceptable in terms of durability, stability, and life span [8]. Comparatively lower energy density of this system can be enhanced by wisely choosing the electrolyte and electrode material. **Figure 1** demonstrates the Ragone plot representing various energy storage systems.

Devices fabricated with TMOs are sophisticatedly important since they exhibit excellent performance in energy storage, wastewater treatment, gas sensing, photovoltaics, etc. In chemical industries, TMOs find application in dye degradation and for the conversion of various hydrocarbons. In energy storage devices, TMO acts as an efficient electrode material especially in supercapacitors and solar cell application. The incomplete d shell corresponding to TMOs resulted in these properties

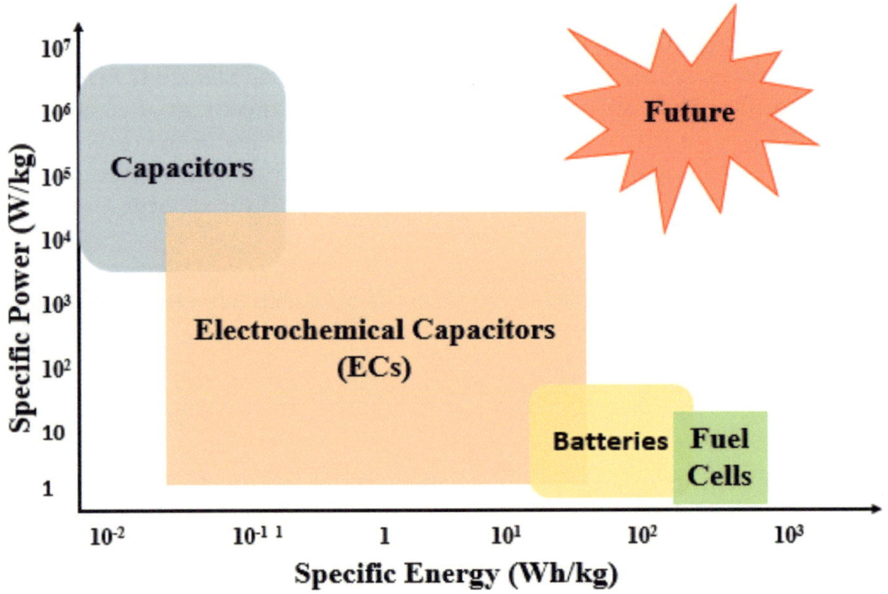

Figure 1.
Ragone plot representing various energy storage systems.

including wide band gap, enhanced chemical reactivity, electrical conductivity, stability, and anti-corrosiveness [9]. To enhance the efficiency of materials, researchers are working on combining TMOs with other transition metals, metal oxides, carbon-based materials, etc. This can modify the surface area, pore characteristics, ion intercalation/deintercalation, conductivity, etc. Many works extending from ZnO/activated carbon to ternary composites such as ZnO/rGO/RuO_2 are still under research study [5].

Graphene, the wonder material with single-atom thickness is one of the most well-liked carbon-based materials because of its high surface area (around 2630 m**2**/g), high conductivity, and chemical stability. The existence of Vander Waals force in graphene results in sudden agglomeration and causes a reduction in surface area and capacity [10]. Carbon-based materials including carbon nanotube (CNT), graphene oxide (GO), activated carbon, etc., are much attracted due to the highly porous structure and enhanced surface area. A combination of TMO with these materials can thus make better electrochemical performance. The size and shape of the nanoparticles can able to tune even the band gap energy of the material; thus, it can control various properties of the nanomaterial including the surface reactivity [11]. These characteristic features of nanostructures can be controlled by various synthesis methods and thus can control the band gap, pore features, etc.

In this chapter, we are focused on the synthesis of some selected TMOs and their composites and discuss the effect of synthesis procedure on the structural and optical characteristics of the material. Also, the article incorporates the device application of the proposed materials.

2. Mechanism of supercapacitors

The reason behind the overview of SCs energy storage system is that SCs weigh less than that of battery with same energy storage capacity, fast access to stored energy, charging very fast than battery, charge/discharge cycle is 106 times, storage capacity is independent of number of charging discharging cycles, negligible environmental concerns, and energy density of SC is 10–100 times larger than that of traditional capacitors [12]. SCs can store substantially more energy than conventional capacitors because the charge separation takes place across a very small distance in the electrical double layer that constitutes the interface between an electrode and the adjacent electrolyte and an increased amount of charge can be stored on the highly extended surface area electrode materials. Electrochemical capacitors also known as supercapacitors exhibit high specific capacitance, high specific power, long cycle life, and fast charge/discharge rate. Theoretical capacitance values of some TMOs are represented in a bar diagram (**Figure 2**).

Supercapacitors are classified according to the energy storage mechanism as electrical double-layer capacitors (EDLCs), pseudocapacitors (PCs), and hybrid capacitors. In EDLCs, charge storage is based on reversible adsorption desorption mechanism at the electrode-electrolyte interface and not involving any faradaic reaction. Electric double layers are formed with the accumulation of charge over the opposite electrodes. Here, the electrode's exposed surface area to the electrolyte determines the capacitance. In other words, pore size of the electrode material should match with the ion size of the electrolyte to avoid capacitance drop. On the other hand, in pseudocapacitors or redox capacitors, charge storage is based on rapid

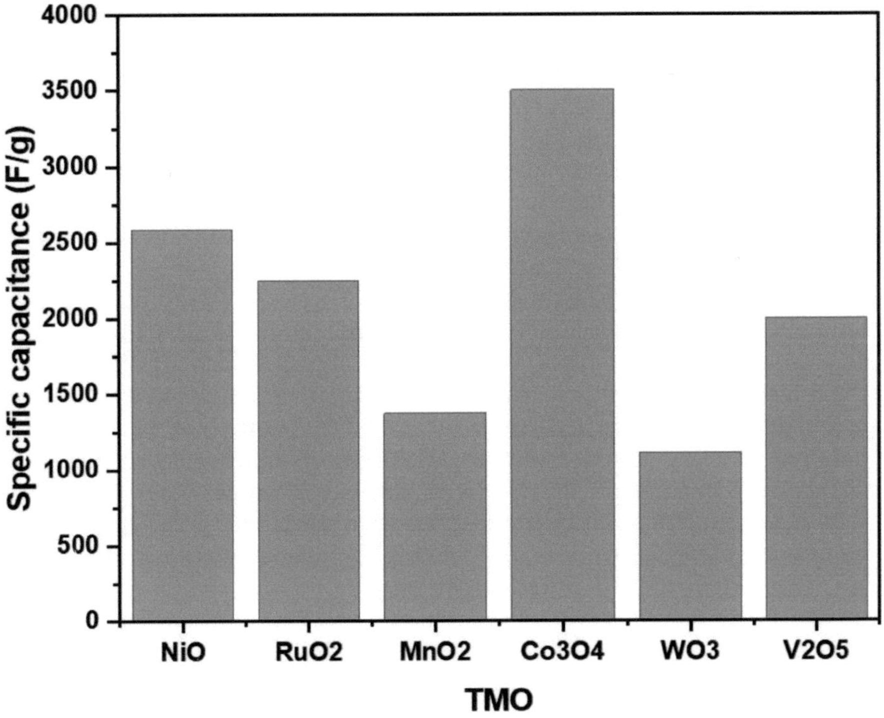

Figure 2.
Theoretical specific capacitance of some TMOs.

faradaic reaction at the surface of electrode material. Generally, pseudocapacitors provide capacitance value higher than EDLCs. In hybrid systems, materials for EDLCs (capacitor-like power sources) and pseudocapacitors (energy sources that resemble batteries) are combined on a single electrode substrate [13].

Electrode is the key factor that determines the performance of any supercapacitor. Depending on the anode and cathode materials in SC device fabrication, it is classified into symmetric, asymmetric, and hybrid device. If two electrodes of the SC are of same material, it is called symmetric device and includes EDLC, pseudocapacitive, and hybrid-type material electrodes. If cathode and anode electrodes are of different types, then the combination is called asymmetric devices. Another classification is based upon the electrolyte. Water-based electrolyte refers as aqueous electrolyte devices, and solvent-based electrolytes come under organic electrolyte-based devices [14].

The electrode material of a supercapacitor is chosen such that it should possess some unique characteristics such as high conductivity, better resistance toward temperature change, large specific surface area, and environmental compatibility. Performance of a supercapacitor relies upon the ability of electrode material for the smooth conduct of faradaic charge transfer [15]. Porosity of the electrode material should be well tuned according to the application where it is used. Small pores yield better surface area, which in turn enhances the specific capacitance and energy density. The better surface area of a porous material enables much more reactive sites and promotes the transfer of electrons and ions. However, this small pore increases the equivalent series resistance (ESR) and hence reduces the specific

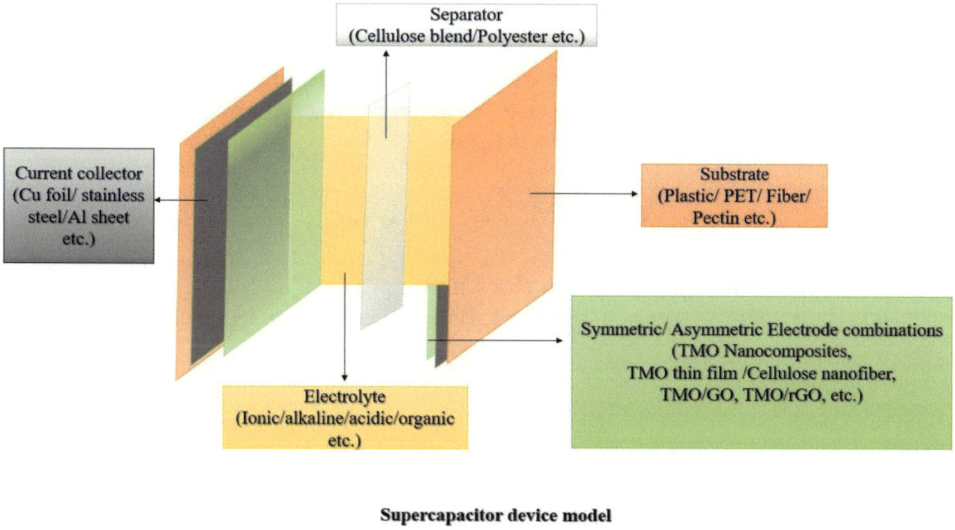

Figure 3.
Schematic representation of supercapacitor device model.

power. Therefore, less porous materials are preferred for applications where high peak current is demanded [16]. Schematic representation of supercapacitor device model is shown in **Figure 3**.

3. Overview of different electrode materials

3.1 Transition metal oxide–based electrode materials

Transition metal oxides (TMOs) are well studied by researchers in the energy storage field. Pseudocapacitive nature of these materials is due to the fast and reversible redox reactions at the surface of electrode material [17]. Theoretical specific capacitance of some TMOs is shown in **Figure 2**.

3.1.1 Manganese dioxide (MnO_2)

Manganese dioxide is well noticed because of high theoretical capacitance (1370 F/g), low cost, and natural abundance. The charge storage mechanism of MnO_2 can be as follows [12]:

$$MnO_2 + \mathrm{M} + +\mathrm{e}^- \leftrightarrow \mathrm{MnOOM} \quad (1)$$

Where M corresponds to electrolyte cations such as K^+, Li^+, Na +, etc. The multiple oxidation states exhibited by MnO_2 result in transition from Mn^{2+} to Mn^{6+} within the potential window. The crystal structure variation by MnO_2 with various polymorphs such as β, α, δ, γ, λ results in variation in electrochemical performance of the material. Other than various advantages, the low conductivity of MnO_2 limits its application as SC electrode material [18].

3.1.2 Vanadium oxide (V_2O_5)

Vanadium pentoxide is an intercalation compound with monovalent cation residing the oxide structure without changing the original structure of the material [14]. The charge storage mechanism is shown as below:

$$V_2O_5 + 2\,M^+ + 2e^- \leftrightarrow M_2V_2O_5 \qquad (2)$$

Layered vanadium oxide structures show better potential for intercalation of diverse ions. Thus, they can be effectively used in LIBs and electrochemical capacitors [19].

3.1.3 Zinc oxide (ZnO)

ZnO is considered as an ideal capacitive material with high energy density of 650 A/g. ZnO is regarded as one of the most suitable materials for pseudocapacitor applications because of its higher electrochemical activity and lower cost. Electric double-layer capacitor (EDLC) nature exhibited by bulk ZnO is replaced by pseudocapacitive behavior by inherently defective nanoscale ZnO. The intrinsic point defects play an important role in device performance as well as energy storage [20]. In this regard, inherently defective ZnO suits well for the pseudocapacitive applications.

3.1.4 Tungsten oxide (WO_3)

Tungsten oxides with WO_6 as the basic octahedra units arranged via sharing corners, edges, or planes, varied with respect to the position of W atoms in the octahedral structure in which the hexagonal phase is reported as one showing best performance.

The charge storage mechanism in WO_3 as follows:

$$WO_3 + M^+ + e^- \leftrightarrow HWO_3 \qquad (3)$$

By different synthesis methods, WO_3 with different size and shape can be synthesized, which includes nanorod, nanopillars, nanosphere, nanoplate, etc. [21]. By controlling the synthesis temperature and reaction time, the morphology as well as the performance of the material can be altered.

3.1.5 Tin oxide (SnO_2)

SnO_2 nanostructures find potential application in the much-needed fields such as energy storage and conversion. This n-type semiconductor holds high electrical conductivity (21.1Ωcm), better theoretical capacity (~782 mAh g^{-1}), low charge-discharge plateau, superior electron mobility (100–200 cm^2/Vs), low synthesis cost, making them a worthy choice for supercapacitor application [22]. Shin et al. [23] obtained a specific capacitance of 40.5 $\mu F/cm^2$ from hierarchical SnO_2 nano branches. They reported a loss of only 8.9% of specific capacitance after 1000 cycles. The large volume change (200, 300%) reported during the charge/discharge process for supercapacitor and Li-ion application leads to pulverization and loss of electrical contact between particles, resulting in low device performance.

3.1.6 Nickel oxide (NiO)

NiO, a p-type semiconductor, has emerged as an efficient electrode material for its higher theoretical capacitance, low toxicity, and environmental impact. Pseudo capacitance property of NiO is greatly affected by its morphology, crystallinity, and conductivity. In general, the pseudo capacitive performance test of NiO is carried out using three-electrode system with nickel foam loaded with active NiO material as the working electrode, platinum sheet and saturated calomel electrode as counter electrode and reference electrode, respectively. A suitable alkaline solution will be selected as the electrolyte. NiO exhibits CV curves like battery materials. Observed redox peaks result from the mutual transformation between Ni (II) and Ni (III). The OH− ions diffused on or inside the electrode participates in the charging and discharging process. The electrochemical reaction can be expressed as [24].

$$NiO + zOH^{-} \leftrightarrow \mathrm{zNiOOH} + (1-\mathrm{z})\,NiO + ze^{-} \tag{4}$$

Observations from various studies record the excellent supercapacitive performance of NiO electrode materials with lower ESR and charge transfer resistance (R_{ct}) [6]. Similarly, in most of the cases, the residual capacitance of NiO electrode materials is more than 90% of initial capacitance even after thousands of charge-discharge cycles.

3.2 TMO/TMO composite

The idea of preparing a composite structure is always encouraged because it imposes a positive result on the overall electrochemical properties of the material. Under the combined effect of various materials/ions, many of the limitations of the transition metal oxides can be surmount [25]. Zheng et al. [26] fabricated a nanostructured electrode material by combining ZnO and NiO. Here ZnO acted as the electrode modifier, and hence, the specific capacitance was improved. Materials capable of exhibiting multiple oxidation states are of particular interest when it comes to electrode fabrication. NiO is one such candidate, whose multiple valence states favor the fast redox reaction, which in turn enhances the specific capacitance. Varshney *et al* investigated the electrochemical properties of nanocomposite of SnO_2 and NiO synthesized by modified sol-gel route. They exhibited a maximum specific capacitance of 464 F/g at a scan rate of 5 mV/s and appreciable capacitance retention of 87.24% after 1000 cycles signifying the fitness of this nanocomposite for high-energy-density supercapacitor electrode material [27].

Multicomponent nanomaterial combination with TMO-TMO heterostructure results in combining both the advantages of individual nanostructures. Tan et al. demonstrated such combination of MnO_2 and V_2O_5 core-shell nanotube by aqueous-based method. It results in a high specific capacitance of 694 F/g at 1 A/g current density with excellent stability [28]. Yu et al. reported the synthesis of $ZnCo_2O_4$ at MnO_2 core-shell nanosheet for asymmetric supercapacitor device fabrication using hydrothermal method. It shows a high specific capacitance of about 2170 F/g at 3 mA/cm^2 in KOH electrolyte. Device fabrication was carried out with TMO heterostructure as the positive electrode and activated carbon as the negative electrode showing an energy density of 29.41 Wh/g with 95.3% retention after 3000 cycles [29].

3.3 TMO/carbon-based composite

Carbon-based materials are generally selected for various applications including supercapacitors, thanks to their high abundance and easy preparation and cost-effectiveness. The porous nature and excellent conductivity make carbon-based materials—graphene, carbon nanotubes, activated carbon, and various other carbon derivatives, a widely accepted electrode materials for supercapacitor applications [30]. As the synthesis techniques are approaching new heights, carbon-based electrodes are generated in various morphologies including nanofibers, nanoflowers, nanorods, nanotubes, etc. MnO_2, V_2O_5, ZnO, and RuO_2 are some of the promising electrode materials.

The performance and structure of electrode material play crucial role in determining the effective capacity of the SC. The limitations of using MnO_2 as electrode material in SC including the poor conductivity can be alleviated by combining with conductive materials such as carbon to make hybrid electrode material. Carbon-based materials such as activated carbon, carbon nanotube, graphene, etc., are attracted due to their pore structure, volume, specific surface area, and presence of functional groups [18]. Cai W et al. proposed an effective method to synthesis of N-doped Carbon@MnO_2 3D core-shell composite, which shows excellent electrochemical performance. The composite shows high specific capacitance with excellent cyclic stability and high retention [31]. Carbon nanotube finds application as electrical double-layer capacitor with high specific surface area and conductivity. But the low specific capacitance limits its usage, thus combined with TMOs. Lei et al. demonstrated a facile method to synthesis MnO_2 nanosheets at graphenated CNTs by hydrothermal method. It shows a high specific capacitance of 575.4 F/g at 0.5 mA/cm^2 and considerably large energy density of about 51.2 Wh/kg [32]. Long et al. reported flexible SC electrode with delta MnO_2 nanosheets anchored on activated carbon cloth. It exhibits a high specific capacitance of 360.5 F/g with capacitive retention of 89.5% after 10,000 cycles [33]. Guo et al. proposed a 3D vertically aligned ZnO nanorods sandwiched between rGO films by chemical vapor deposition. Specific capacitance of 51.6 F/g is achieved at 10 mV/s for supercapacitor performance [34]. Poor electrical conductivity of V_2O_5 can be balanced by the proper combination with conducting materials as that proposed by Perera et al. CNT/V_2O_5 nanowire nanocomposite thus exhibits an ideal capacitive behavior with specific capacity of 48.5 F/g and power density of 5.26 kW/kg [35].

Compared with all other carbon-based materials, graphene has attracted tremendous attention in high-performance energy storage systems because of its intriguing properties including large surface area, excellent conductivity, commendable thermal, optical, and mechanical properties. Basically, graphene is a single layer of atom constructed by sp2-bonded carbon atoms arranged in a poly aromatic honey comb crystal structure. Theoretical specific capacitance of graphene-based EDLC is 550 Fg^{-1}. In literature, one can see that supercapacitors based on graphene exhibited a specific capacitance of 75 Fg^{-1} with an energy density of 31.9 $Whkg^{-1}$ in ionic liquid electrolytes and 135 Fg^{-1} specific capacitance in aqueous electrolyte. On the other hand, it showed a specific capacitance of 99 Fg^{-1} in organic electrolytes [36]. But it is noted that the restacking of graphene sheets reduces its conductivity and results in poor specific capacitance. Restacking occurs due to the van der Waals interaction between the sheets, and it diminishes coulombic efficiency also. In order to improve the capacitive nature of graphene, they are usually made composites with other capacitive materials. Graphene-metal oxide composite seems to be a good combination since metal oxide hinders graphene from restacking. Metal oxides carry the role of a stabilizer, which prevents the accumulation of graphene sheets. Combination of graphene and metal oxide compliments each other by eliminating the complications faced by these materials individually. Development of composites with pseudocapacitive materials possesses an advantage of generating capacitance from redox charge transfer in addition to the double-layer capacitance. In metal oxide-graphene composite, graphene acts as a passage for charge transfer, whereas metal oxides provide pseudo capacitance. ZnO is a versatile material possessing 3.37 eV bandgap, with an exciton binding energy of 60 meV, which makes it suitable for a wide range of applications including supercapacitors. Regardless, the lower specific capacitance exhibited by ZnO compared with other metal oxides, ZnO superiors with low cost, high abundance, and less toxicity. Also, the electron donating nature of ZnO makes it a good partner for electron acceptor graphene to make efficient electrode materials. As reported by Dutta et al. [37], ZnO/rGO composite electrode can achieve a high specific capacitance of 1012 F/g at a current density of 1 A/g with an outstanding power density of 3534.6 W/kg. Sreejesh et al. [38] further prepared ZnO/rGO nanocomposite by microwave-assisted technique to achieve a high capacitance up to 631 F/g and a long life cycle tested up to 2000 cycles. Introduction of graphene can effectively transcend the poor electrical conductivity of NiO, another promising transition metal oxide candidate. The core-shell hybrid NiO/rGO electrode prepared by electrophoretic deposition method shows a capacitance of 940 F g^{-1} at a current density of 2 A g^{-1} [39]. Pore et al. reported the achievement of specific capacitance of 727.1 F g^{-1} at 1 mA cm^{-2} current density with good cyclic stability of about 80.4% over 9000 cycles for the hydrothermally prepared NiO/rGO electrodes [40].

3.4 TMO/TMD nanocomposite

In energy storage application, the limitations in 2D materials as potential electrodes for energy storage include the graphene-based electrodes having high electrical conductivity and mechanical strength, but it demonstrates only moderate capacity due to the charge storage on the surface only, thus decreasing its conductivity. Transition metal dichalcogenide (TMDs) shows high initial capacities, but undergoes a conversion reaction on the first discharge cycle leading to poor capacity retention. Some TMDs demonstrate relatively high electronic conductivity; many of the phases are semiconducting [41]. Transition metal oxides (TMOs) show high redox activity in intercalation reactions and relatively high working potentials making them especially attractive for use as electrodes [42]. However, low electronic conductivity of oxides imposes a requirement to mix them with a conductive additive to improve their performance. Electrochemical properties of MXenes strongly depend on the synthesis conditions and surface chemistry, and methods for control of their surface terminations need to be developed to minimize irreversible capacity. Thus, research studies point toward an active material development, construction of low cost, large scale, eco-friendly production of TMD/TMO heterostructure. To overcome the limitations of individual 2D materials such as irreversible capacity loss due to consumption of large amounts of electrolyte in the solid-electrolyte interface, electrolyte decomposition due to large no of active sites for ions intercalation, it can be combined with high redox activity of TMO. The heterostructure can create an abundance of structural defects and multiple accessible electrochemically active sites for ion/electron migrations, which would largely enhance the redox reaction activities toward electrochemical energy storages [43]. A rough schematic diagram involving fabrication of TMO/TMD heterostructure is shown in **Figure 4**.

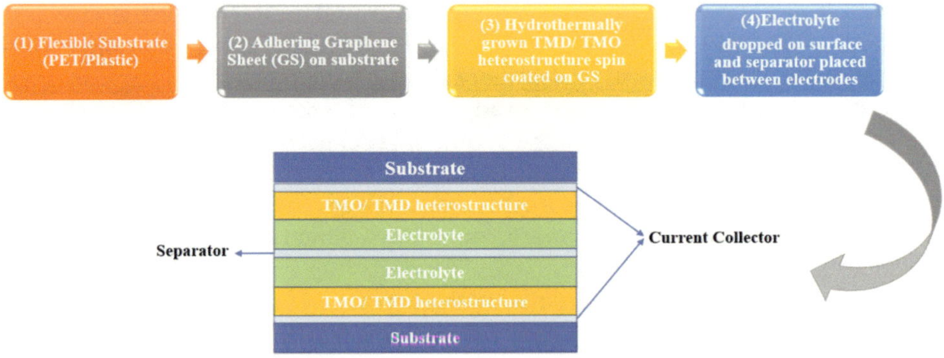

Figure 4.
The fabrication process in TMO/TMD heterostructure-based supercapacitor.

Kai Wang and coworkers [44] focused on general solution-processed formation of porous transition-metal oxides on exfoliated molybdenum disulfides for high-performance asymmetric supercapacitors. They prepared few-layered MoS_2 (f- MoS_2) with a series of TMOs (Ni, Co, and Fe-based oxides) via a chemical bath deposition method at ambient conditions. Each combination of MoS_2 with the abovementioned TMO is synthesized and compared its electrochemical performances. They assembled asymmetric supercapacitor electrode device with a MoS_2-NiO//MoS_2-Fe_2O_3 configuration, and the device shows a maximum energy density and good cycling performance. K. Chanda and coworkers [45] prepared 3D hierarchical nanoform based on TiO_2 sphere and MoS_2 flake by low-temperature hydrothermal synthesis protocols. Surface area increased due to the wrapping of MoS_2 flakes over TiO_2 sphere. Hydrothermal treatment of these spheres creates a drastic change in their morphology. The electrochemical performance of the hybrid is compared with pristine TiO_2 and found the improvement in capacitance values.

3.5 TMO/conducting polymer for SC

Conducting polymers are pseudocapacitive materials, which store charge under fast redox reaction at the surface and bulk part of electrode material. CPs having good intrinsic conductivity show poor cyclic stability due to the variation observed in polymer chain during the doping-dedoping process. It is a low-cost material with excellent charge density. It includes polyaniline (PANI), polypyrrole (PPy), and poly (3,4- ethylenedioxythiophene) (PEDOT) having theoretical capacitance of about 100 F/g. The specific capacitance shown by CPs is two times greater than that of EDLCs. On the other hand, TMOs having low conductivity limit its applicability to achieve the high theoretical capacitance. By combining the advantages of both, the performance might be enhanced with high specific capacitance and good electrical conductivity compared with pure ones [46] without any loss in the pseudocapacitance of the material.

4. Summary

Supercapacitors are currently a new class of energy storage devices that have applications in many different industries. The development of novel, effective electrode materials is a global priority. Transition metal oxides are a desirable material for supercapacitors due to their ideal capacitive performance and much lower cost and environmental compatibility. They store charge in a pseudocapacitive manner. The primary properties of TMO are its significant inherent stability and its challenging valence, which allow for the intercalation of electrons and ions into the lattice of metallic compounds. For high-performance supercapacitors that can give high capacitance, higher cyclic stability, and exceptional rate, transition metal oxides, carbonaceous electrode materials, and their composites considered in this chapter look promising.

Author details

Nithya S. George, Lolly Maria Jose and Arun Aravind*
Center for Advanced Functional Materials, Department of Physics, Bishop Moore College, Mavelikara, India

*Address all correspondence to: arun@bishopmoorecollege.org

IntechOpen

© 2022 The Author(s). Licensee IntechOpen. This chapter is distributed under the terms of the Creative Commons Attribution License (http://creativecommons.org/licenses/by/3.0), which permits unrestricted use, distribution, and reproduction in any medium, provided the original work is properly cited.

References

[1] Kulandaivalu S, Hussein MZ, Jaafar AM, Abdah MA, Azman NH, Sulaiman Y. A simple strategy to prepare a layer-by-layer assembled composite of Ni–Co LDHs on polypyrrole/rGO for a high specific capacitance supercapacitor. RSC Advances. 2019;**9**(69):40478-40486

[2] Liang R et al. Transition metal oxide electrode materials for supercapacitors: A review of recent developments. Nanomaterials. 2021;**11**:1248. DOI: 10.3390/nano11051248

[3] Ali S, Saleh L, Hadi R, Farhoudian S, Nedaei M. Transition metal oxide-based electrode materials for flexible supercapacitors: A review. 2020;**857**

[4] Cui M, Meng X. Overview of transition metal-based composite materials for supercapacitor electrodes. Nanoscale Advances. 2020;**2**:5516-5528. DOI: 10.1039/ DONA00573H

[5] Low WH, Khiew PS, Lim SS, Siong CW, Ezeigwe ER. Recent development of mixed transition metal oxide and graphene/mixed transition metal oxide-based hybrid nanostructures for advanced supercapacitors. Journal of Alloys and Compounds. 2019;**775**:1324-1356. DOI: 10.1016/j.jallcom.2018.10.102

[6] Maduraiveerana G, Sasidharan M, Jin W. Earth-abundant transition metal and metal oxide nanomaterials: Synthesis and electrochemical applications. Progress in Materials Science. 2019;**106**:100574. DOI: 10.1016/j.pmatsci.2019.100574

[7] Housel LM, Takeuchi ES. Investigation of alpha-MnO2 tunneled structures as model cation hosts for energy storage. 2018;**51**(3):575-582. DOI: 10.1021/acs.accounts.7b00478

[8] Sivakumar S, Nelson Prabu L. Synthesis and characterization of a-MnO2 nanoparticles for supercapacitor application. Materials Today: Proceedings. 2021;**47**:52-55. DOI: 10.1016/j.matpr.2021.03.528

[9] Kung HH. Transition Metal Oxides: Surface Chemistry and Catalysis1989. p. 45 ISBN 0-444-87394-5

[10] Zhang Y et al. A green hydrothermal approach for the preparation of graphene/alpha -MnO2 3D network as anode for lithium ion battery. Electrochimica Acta. 2013;**108**:465-471

[11] Dong et al. MnO2/rGO/CNTs framework as a sulfur host for high-performance Li-S batteries. Molecules. 2020;**25**:1989. DOI: 10.3390/molecules25081989

[12] Lokhande CD, Dubal DP, Joo O-S. Metal oxide thin film based supercapacitors. Current Applied Physics. 2011;**11**:255-270

[13] Pushparaj VL, Shaijumon MM, Kumar A, Murugesan S, Ci L, Vajtai R, et al. Flexible energy storage devices based on nanocomposite paper. Proceedings of the National Academy of Sciences. 2007;**104**(34):13574-13577. DOI: 10.1073/pnas.0706508104

[14] Shivaji Pawar, Progress and Prospects in Nanoscience Today. Unites states: Novan Science Publishers, Inc.; 2020. ISBN: 978-1-53617-293-5

[15] Ratha S, Samantara AK. Supercapacitor: Instrumentation, Measurement and Performance Evaluation Techniques. Singapore: Springer; 2018. ISBN 978-981-13-3085-8. DOI: 10.1007/978-981-13-3086-5

[16] Rajagopal S, Pulapparambil Vallikkattil R, Mohamed Ibrahim M, Velev DG. Electrode materials for supercapacitors in hybrid electric vehicles: Challenges and current Progress. Condens. Matter. 2022;7(1):6. DOI: 10.3390/condmat7010006

[17] Ray A, Roy A, Saha S, Das S. Transition metal oxide-based Nano-materials for energy storage application. Science, Technology and Advanced Application of Supercapacitors. London, United Kingdom: Intechopen. 2018 [Online]. DOI: 10.5772/intechopen.80298

[18] Noh J, Yoon CM, Kim YK, Jang J. High performance asymmetric supercapacitor twisted from carbon fiber/MnO2 and carbon fiber/MoO3. Carbon. 2017;**116**:470-478.1

[19] Delbari SA, Ghadimi LS, Hadi R, Farhoudian S, Nedaei M, Babapoor A, et al. Transition metal oxide-based electrode materials for flexible supercapacitors: A review. Journal of Alloys and Compounds. 2021;**857**:15828

[20] Shaheen I, Ahmad KS, Zequine C, Gupta RK, Thomas AG, Malik MA. Facile ZnO-based nanomaterial and its fabrication as a supercapacitor electrode: Synthesis, characterization and electrochemical studies. RSC Advances. 2021;**11**(38):23374-23384. DOI: 10.1039/d1ra04341b

[21] Zheng F, Xi C, Xu J, Yu Y, Yang W, Hu P, et al. Facile preparation of WO3 nano-fibers with super large aspect ratio for high performance supercapacitor. Journal of Alloys and Compounds. 2019;**772**:933-942. DOI: 10.1016/j.jallcom.2018.09.085

[22] Venkataramana B, Bhavana G, Sharat C, Arindam D, Sandip D, Tyagi AK. Electrochemical supercapacitor performance of SnO2 QDs. Electrochimica Acta. 2016;**203**:230-237

[23] Shin JH, Park HM, Song JY. Phase transformation of hierarchical nanobranch structure from SnO to SnO2 and its electrochemical capacitance. Journal of Alloys and Compounds. 2013;**551**:451

[24] Zhao J, Tian Y, Liu A, Song L, Zhao Z. The NiO electrode materials in electrochemical capacitor: A review. Materials Science in Semiconductor Processing. 2019;**96**:78-90

[25] Kamal K. Kar, Handbook of Nanocomposite Supercapacitor Materials I Characteristics. Cham: Springer, Springer Series in Materials Science; 2020. ISBN 978-3-030- 43008-5. DOI: 10.1007/978-3-030-43009-2

[26] Zheng JH, Zhang RM, Wang XG, Yu PF. Synthesizing a flower-like NiO and ZnO composite for supercapacitor applications. Research on Chemical Intermediates. 2018;**44**(9):5569-5582

[27] Varshney B, Siddiqui MJ, Anwer AH, Khan MZ, Ahmed F, Aljaafari A, et al. Synthesis of mesoporous SnO2/NiO nanocomposite using modified sol–gel method and its electrochemical performance as electrode material for supercapacitors. Scientific Reports. 2020;**10**(1):13

[28] Tan HT, Rui X, Yu H, Liu W, Xu C, Xu Z, et al. Aqueous-based chemical route toward ambient preparation of multicomponent core–shell nanotubes. ACS Nano. 2014;**8**(4):4004-4014. DOI: 10.1021/nn500942k

[29] Yu D, Zhang Z, Meng Y, Teng Y, Wu Y, Zhang X, et al. The synthesis of hierarchical ZnCo2O4@MnO2 core–shell nanosheet arrays on Ni foam

for high- performance all-solid-state asymmetric supercapacitors. Inorganic Chemistry Frontiers. 2018;5(3):597-604. DOI: 10.1039/c7qi00706j

[30] Du YQ, Xiao P, Yuan J, Chen JW. Research Progress of graphene-based materials on flexible supercapacitors. Coatings. 2020;**10**:892

[31] Cai W, Kumar Kankala R, Xiao M, Zhang N, Zhang X. Three-dimensional hollow N-doped ZIF-8-derived carbon@ MnO2 composites for supercapacitors. Applied Surface Science. 2020:146921. DOI: 10.1016/j.apsusc.2020.146921

[32] Ye L, Wang C, Le Y, et al. Construction of MnO2 nanosheets @graphenated carbon nanotube networks core-shell heterostructure on 316L stainless steel as binder-free supercapacitor electrodes. International Journal of Hydrogen Energy. 2020;**45**(53):28930-28939. DOI: 10.1016/j.ijhydene.2019.09.070

[33] Long X, Tian L, Zhang L, Wang J, Chen Y, Emin A, et al. Interconnected δ-MnO2 Nanosheets anchored on activated carbon cloth as flexible electrode for high-performance aqueous asymmetric supercapacitors. Journal of Electroanalytical Chemistry. 2020;**877**:114656

[34] Guo G, Huang L, Chang Q, Ji L, Liu Y, Xie Y, et al. Sandwiched nanoarchitecture of reduced graphene oxide/ZnO nanorods/reduced graphene oxide on flexible PET substrate for supercapacitor. Applied Physics Letters. 2011;**99**(8):083111. DOI: 10.1063/1.3629789

[35] Perera SD, Patel B, Nijem N, Roodenko K, Seitz O, Ferraris JP, et al. Vanadium oxide nanowire-carbon nanotube binder-free flexible electrodes for supercapacitors. Advanced Energy Materials. 2011;**1**(5):936-945. DOI: 10.1002/aenm.201100221

[36] Gilmore KJ, Wallace GG, Li D, Chen BH, Mu MB. Mechanically strong, electrically conductive, and biocompatible graphene paper

[37] Dutta A, Mishra S, Saha SK, Sarkar S, Guchhait A, Akhtar AJ. Boosting the Supercapacitive performance of ZnO by 3-dimensional conductive wrapping with graphene sheet. Journal of Inorganic and Organometallic Polymers and Materials. 2022;**32**(1):180-190

[38] Sreejesh M, Dhanush S, Rossignol F, Nagaraja HS. Microwave assisted synthesis of rGO/ZnO composites for non-enzymatic glucose sensing and supercapacitor applications. Ceramics International. 2017;**43**:4895-4903

[39] Yus J, Bravo Y, Sanchez-Herencia AJ, Ferrari B, Gonzalez Z. Electrophoretic deposition of RGO-NiO core-shell nanostructures driven by heterocoagulation method with high electrochemical performance. Electrochimica Acta. 2019;**308**:363-372

[40] Pore OC, Fulari AV, Parale VG, Park HH, Shejwal RV, Fulari VJ, et al. Facile hydrothermal synthesis of NiO/rGO nanocomposite electrodes for supercapacitor and nonenzymatic glucose biosensing application. Journal of Porous Materials. 2022:1-1

[41] Liang X, Zhang X, Liu W, Tang D, Zhang B, Ji G. A simple hydrothermal process to grow MoS2 nanosheets with excellent dielectric loss and microwave absorption performance. Journal of Materials Chemistry C. 2016;**4**(28):6816-6821. DOI: 10.1039/C6TC02006B

[42] Kanaujiya N, Nagesh Kumar AK, Srivastava YS, Varma GD. One-step synthesized mesoporous MnO2@

MoS2 nanocomposite for high performance energy storage devices. Journal of Electroanalytical Chemistry. 2018;**824**:226-237. DOI: 10.1016/j.jelechem.2018.07.046

[43] Prabhu P, Jose V, Lee J-M. Design strategies for development of TMD-based Heterostructures in electrochemical energy systems. Matter. 2020;**2**:526-553. DOI: 10.1016/j.matt.2020.01.001

[44] Wang K, Yang J, Zhu J, Li L, Liu Y, Zhang C, et al. General solution-processed formation of porous transition-metal oxides on exfoliated molybdenum disulfides for high-performance asymmetric supercapacitors. Journal of Materials Chemistry A. 2017;**5**(22):11236-11245. DOI: 10.1039/c7ta01457k

[45] Chanda K, Thakur S, Maiti S, Acharya A, Paul T, Besra N, et al. Hierarchical heterostructure of MoS2 flake anchored on TiO2 sphere for supercapacitor application. AIP Conference Proceedings. 2018;**1953**(1):030138. DOI: 10.1063/1.5032473

[46] Abdah MAAM, Azman NHN, Kulandaivalu S, et al. Review of the use of transition-metal-oxide and conducting polymer-based fibres for high-performance supercapacitors. Materials & Design. 2019;**186**:108199. DOI: 10.1016/j.matdes.2019.108199

Chapter 5

Supercapacitors: Fabrication Challenges and Trends

Yi Su and Mohamad Sawan

Abstract

Supercapacitors have shown great potential as important complements to batteries. We first describe the principle of supercapacitors, including the categories and the main components of supercapacitors. In the second part, we compare the advantages of supercapacitors with other energy storage devices, and then the power densities of active materials are compared with each other. In the third part, we show how various technologies are used to fabricate electrodes and supercapacitors. In the last part, several applications are presented, showing the high value of supercapacitors, including hybrid vehicles, solar cells, and wearable and portable devices.

Keywords: supercapacitors, energy storage, portable devices, wearable devices

1. Introduction

The urgent need for renewable energy sources has resulted in concerted research efforts into electrochemical energy storage. Capacitors can maintain power for a long time, according to the following equation:

$$E = \tfrac{1}{2} C V^2 \tag{1}$$

where C is the capacitance and V is the applied voltage [1]. However, even though capacitors have been known to scientists for many centuries, conventional capacitors still have values at pF and uF/cm^2, which are far from energy storage requirements. When applying capacitors to energy storage, the capacity of capacitors must be increased, which are known as supercapacitors. Following developments over the past decade, supercapacitors have shown great promise for next-generation energy storage devices [2]. Supercapacitors feature a high power density [3], fast charge and discharge capacity, environmental friendliness [4], safety, and a long life cycle [5], allowing them to be used as important complements to batteries [6]. To promote the application field of supercapacitors for researchers, in this chapter, we introduce the main characteristics of supercapacitors, including categories, components, advantages, fabrication, and applications. Moreover, we discuss future application trends.

In the first section, we introduce the types and main components of supercapacitors, which can be separated into three categories: (1) double-layer capacitors [7], (2) pseudo-capacitors, and (3) hybrid supercapacitors [8], which consist of active materials, collectors, separators, and electrolytes. Conductive metal-organic frameworks operate as collectors, while the active materials are the main components that determine the power density of supercapacitors. Separators are composed of fibers and polymers, while ions are the main components of the electrolytes. Moreover, we illustrate the structure of supercapacitors and compare the power densities in different active materials. **Table 1** shows the various supercapacitors and their power densities. The MnO_2 and NiO are the most popular oxide metals with the high energy density, and they usually worked with the graphene and carbon nanotubes (CNTs) to fabricate the hybrid supercapacitors. As can be seen, both the flexibility and stretch ability are common parameters in advanced supercapacitors; however, the high energy density hardly achieved in the supercapacitors compared to the Li-batteries.

Comparing with other energy storages, this chapter exhibits many advantages, including the long-term cycling stability, high safety, and power density, which make them promising candidates for energy storage in many applications.

No	Active materials	Parameters	Capacitance	Applied voltage	Reference
1	AU/MnO_2	Transparency, flexibility	524 F/g	2 V	[9]
2	AU/MnO_2	Transparency, flexibility	1370 F/g	1.5 V	[10]
3	CNTs/PANI	Flexibility	243 F/g	0.8 V	[11]
4	AU/MnO_2	Transparency, flexibility	3.68 mF/cm^2	1 V	[12]
5	Ni-Co oxide/ MnO_2	Stretchability	0.77 F/cm^2	1 V	[13]
6	NiO/carbon	N/A	317 F/g	0.5 V	[14]
7	rGO-PPY	Flexibility	232 mF/cm^2	0.8 V	[15]
8	Nitrogen-doped graphene	Stretchability	806 F/cm^3	1 V	[16]
9	GO@PPY	Flexibility	419 mF/cm^2	1 V	[17]
10	Self-doped lignin-based biocarbon	Flexibility	140 F/g	1 V	[18]
11	Multi-walled CNTs	N/A	671 F/g	0.7 V	[19]
12	PPY@rGO	N/A	882.2 F/g	1 V	[20]
13	GO@carbon aerogel@MnO_2	Micro	8.7 mF/cm^2	1 V	[21]
14	PEDOT: PSS	Flexibility, deformability	232 mF/cm^2	1 V	[22]
15	Multi-walled CNTs	Flexibility, stretchability, and self-charging	167 mF/cm^2	0.9 V	[23]
16	PPY	Stretchability and micro-characteristics	221.2 mF/cm^2	0.6 V	[24]
17	PPY	Flexibility	203 mF/cm^2	0.8 V	[25]

No	Active materials	Parameters	Capacitance	Applied voltage	Reference
18	GO@-MnO_2	Micro	1100 F/cm^3	1 V	[26]
19	CNTs	Flexibility/ stretchability	0.6 mF cm^{-2}	2.5 V	[7]
20	PPY@graphene	Flexibility	89.6 mF/cm^2	1.4 V	[27]
20	PPY	Transparency	5.6 mF/cm^2	0.6 V	[28]
21	Ag/NixFeyOz@ rGO	Transparency	282.1 uF/cm^2	4 V	[8]
22	Carbon fiber	Flexibility	246.1 F/g	1.8 V	[29]
23	PPY	Micro	128 mF/cm^2	1 V	[30]
24	GO/CNTs	Flexibility	269 mF/cm^2	2.5 V	[31]
25	GO	Environmentally friendly	173.8 F/g	1 V	[4]
26	ZIF-L (Zn)@ $Ti_3C_2T_x$	Flexibility	854 F/cm^3	1 V	[32]
27	MnO_2@TiN NWs@CNTs	Flexibility	61.2 mW h cm^{-3}	3.5 V	[3]
28	NiO@graphene	Micro- characteristics	605.9 mF/cm^2	0.8 V	[33]
29	PANI/N-CNTs@ CNTs	Flexibility	323.8 F/g	0.8 V	[34]

Table 1.
Parameters of various main supercapacitors.

In this chapter, we present the advanced achievements of supercapacitors over the past 5 years, and we overview the technologies used for fabricating supercapacitors, including lithography, laser writing, inkjet printing, template sacrifice, and physical and chemical vapor deposition.

We also present an overview of applications where supercapacitors have been widely used such as in vehicles and other storage buffers. We describe the stretchable and flexible supercapacitors in wearable devices. We also present wirelessly rechargeable flexible supercapacitors used in soft and smart lenses. The later have shown potential applications for implantable and wearable medical devices. Moreover, we present supercapacitors that may be completely biodegradable and bioabsorbable, which has never been achieved in other power storage devices. As complementary storage for batteries, supercapacitors are becoming increasingly important in several applications.

PANI, polyaniline; GO, graphene Oxide; PPY, polypyrrole; CNTs, carbon nanotubes; PEDOT, poly(3,4-ethylenedioxythiophene); PSS, polystyrene sulfonate.

2. The categories of supercapacitors and main components of supercapacitors

There are three types of categories (**Figure 1**), depending on the charge storage mechanism. (1) The first one is double-layer capacitors [35], which form capacitance using double large electrodes, consisting of a physical process. The sandwich structure forms capacitance, according to the following equation:

$$C = \varepsilon s / d, \tag{2}$$

where C is the capacitance and is determined by ε, s, and d, where s is the area of the two electrodes, d is the thickness between them, and ε is the permittivity of the electrolyte. The capacitance will also be different between capacitors and supercapacitors. The active materials consist of CNT, graphene, and a conductive polymer with a large area that is stable during the charge and discharge process. (2) For pseudo-capacitors [36], the energy storage in pseudo-capacitive comes from the surface redox reaction [2] between the metal oxide and conductive polymers, as they generally have many oxidation states that can absorb and emit electrons such as Ni oxide, Co oxide, Mn oxide, and Co oxide. (3) Hybrid supercapacitors, which mix double-layer capacitors and pseudo-capacitors, have more advantages than each individual type. They exhibit a high charge ability and an increase in the capacitance compared with double-layer capacitors, resulting in good performance with high cycle and charge abilities, with the same high capacitance as pseudocapacitors. Based on the discussion mentioned earlier, the hybrid supercapacitors are future trend to overcome the limitation of power density and applications. The supercapacitors can be divided into different categories, and they all have the same components, including collectors, separators, electrolytes, and active materials, with designed pore space for active materials.

2.1 Collectors

Conductive metal-organic frameworks operate as collectors and can be regarded as the supporting components for capacitors. To fabricate metal-organic frameworks, various approaches can be used, which can be divided into bare metal foam, sacrificial templates, micro/neon network structures with in situ deposition, and free-standing fiber films through electrospinning. Cellulose and self-doped multi-porous lignin-based biocarbon with a three-dimensional network structure are frameworks that can be used as the substrate of a solid-state flexible supercapacitor. The shape design of anodic aluminum oxide (AAO) and nanomesh template [37] have shown excellent nanostructures and mechanical properties, making them suitable for the fabrication of supercapacitors with many properties, exhibiting easy fabrication, versatility, and a high surface area. Sacrificial polystyrene colloidal particles [19] and cube sugar were shown to increase the power intensity with a large specific surface area per mass volume for the electrodes. The metal coating on the pre-electrospinning fiber film resulted in the formation of metal-framework networks, with a process to fabricate a larger area collector with auto-fabrication. Other approaches have also

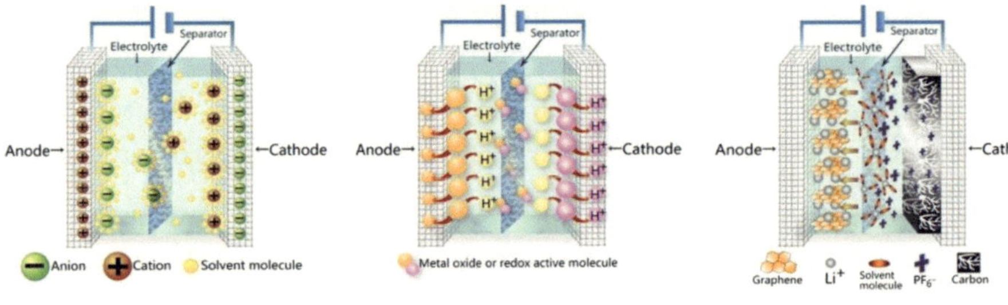

Figure 1.
Three categories of supercapacitors and they consist of collectors, active materials, separators and electrolytes.

been used, such as wrinkled graphene on polydimethylsiloxane [7], as well as crackle and leaf templates, which have been used to fabricate metal frameworks with deposited conductive metals. The shape design of collectors is more suitable used in various fabrications and applications due to its higher energy density.

2.2 Separators and electrolyte

Separators are filters through which ions migrate in and out, and they can reject connections between two electrodes. Resistance and stability are the key parameters for separators. Many types of commercial separators are available, which consist of fiber separators, high polymer separators, electrospinning separators, and biological separators. The poly(vinyl alcohol) (PVA), which is a high polymer separator, has shown high conductivity, biodegradability, and high alkaline stability. PVA-based gels in aqueous electrolytes have been researched for solid-state supercapacitor applications and fuel cells. However, PVA-based gels have some limitations in aqueous solutions, i.e. the voltage window should be under or equal to 1 V. Researchers have now focused on finding inorganic ceramic solid-state electrolytes. Polyacrylonitrile (PAN) is a type of electrospinning separator that has shown great shrinkage and porosity, with many applications. In addition, studies have been conducted on biological separators, such as egg shell membranes [38], fish bubbles [39], and other biological separators. Furthermore, separators should be stretchable, stable, and in a staining state in flexible and stretchable supercapacitors, which has shown to be a challenging property in separators. Electrolytes consist of a conductive liquid mixture containing an aqueous or organic solvent, based on active materials and ions that include OH^-, H^+, Li^+, Na^+, K^+, SO_4^{2-}, and Cl^- [40]. The concentration of ions and the PH value have shown the impact of power intensity in supercapacitors, and the less volume ions performed better in the transfers with less impedance.

2.3 Active materials

Depending on the charging mechanism, active materials can usually be divided into four categories. (1) Carbonaceous materials include CNTs, graphene, and its allotrope, where graphene is mainly based on double-layer capacitance, which utilizes only the surface of the electrode material; however, the capacitance of bare carbonaceous will be relatively small, limiting the applications of supercapacitors. (2) Metal oxides, especially Ni oxide, Co oxide, Mn oxide, and Co oxide, include MnO_2 [12], NiO, Co_3O_4, and $Co(OH)_2$ [13], and these metals have many energy levels [41], which can easily absorb and emit some electrons. Furthermore, the power density of these metal oxides is the largest among all active materials; however, they have shown poor high cycle and charge abilities, as well as environmental pollution. (3) Conductive polymers include polypyrrole (PPY) [24], polyaniline (PANI) [11], polythiophenes (PYH), polyphenylenevinylene (PPA), and polyacetylene (PA). These conductive polymers have shown good mechanical properties, easy fabrication, power density, and environmental friendliness. (4) Composites of the aforementioned active materials have shown better performance in all of these parameters, especially a higher power density, better mechanical properties, and charge and cycle abilities. These materials include graphene@ MnO_2, PPY@NiO, Graphene@PPY film, C@ TiN [3], and ZIF-Li(Zn)@TisC2Tx [32]. Comparing the performances of the single active materials, the association of the different active materials shows the promising candidates in power storages.

2.4 Designed pore space for active materials

The power density relies both on the active materials and pore space of the active materials. The optimization of the structure of active materials and dynamics of ions at

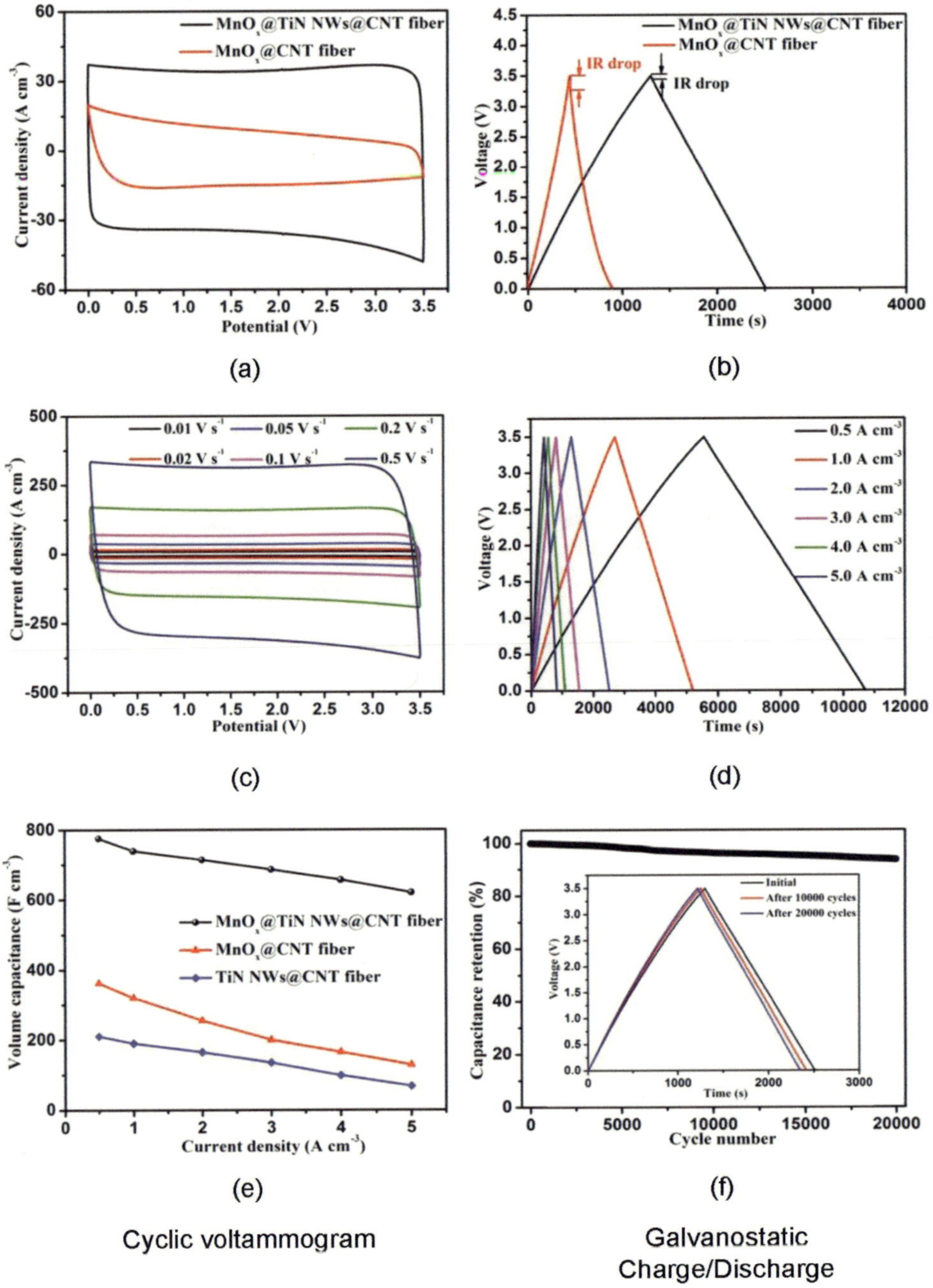

Figure 2.
(a) and (b) CV curves and GCD curves of two different fiber electrodes, (c) CV curves under different scan rates, (d) GCD curves at different current densities of the MnOx@TiN NWs@CNT fiber electrode, (e) comparison of volumetric capacitances versus different current densities for three different fiber electrodes, and (f) cycling stability of the MnOx@TiN NWs@CNT fiber electrodes as a function of cycle number under a current density of 2.0 A cm^{-3} *[3].*

the electrode-electrolyte interface are the two promising routes for enhancing the electrochemical performance of supercapacitors [2]. The immobilized ions diffuse into the electrolyte, which not only is the key parameter for the supercapacitors, but also affect the impedance of the supercapacitors. For controlling the diffusing rate, researchers have designed and fabricated various superior architectures, including fabricating 3D structures [27], which include MXenes structures, core-sheath fibers, asymmetric fiber structures [3], and hierarchical nanoarrays. The synthetic strategy of hierarchical nanoarrays [33] can be used as a template for the epitaxial growth of secondary structures by controlling the reaction condition using a one-pot strategy to fabricate the hierarchical structure. Studies have shown that three-dimensional structures will increase the surface area considerably, especially in shape design structures. The asymmetric fiber structure has shown high potential applications because of multidimensional stacking and winding, resulting in unbelievable capacitance and power density values. **Figure 2** shows the different parameters of the supercapacitors, and **Figure 2a** and **c** shows the stability performance in the cyclic voltammetry curves at different scan rates. The charge and discharge curves of the different supercapacitors at different current densities are shown in **Figure 2b** and **d**, where the volumetric capacitances were calculated from the charge and discharge rates as a function of the current density in different capacitances (**Figure 2e**). **Figure 2f** shows the stability of supercapacitors, indicating the performance of the supercapacitors under different current and voltage values, which can be used to design the pore space and select the suitable active material.

3. Supercapacitors offer many advantages compared with other energy storage devices

Traditionally, batteries are the main energy storage centers in devices, with a hard shell and the toxicity of the electrolyte in coin cells. Compared with lithium-ion batteries, the cell configurations of supercapacitors will be much simpler and safer due to the use of water or polymer gel electrolytes instead of flammable and moisture-sensitive organic electrolytes. Furthermore, the fast and highly reversible double-layer capacitive and/or pseudocapacitive energy storage mechanism of supercapacitors will enable a significantly better high-power capability and cycling stability compared with lithium-ion batteries. Even if the lower energy density of supercapacitors restricts the wide application of supercapacitors, due to their superior high-power performance, long-term cycling stability, and high safety, supercapacitors have been regarded as promising candidates for energy storage in some applications, as introduced in the following section.

3.1 High safety

Compared with ion batteries, which use toxic electrolytes with a hard shell, many biocompatible electrodes can be used in supercapacitors, with a more moderate electrolyte. Some supercapacitors can exhibit self-degradation, biodegradation, and bioabsorption [42] with no effects on the body, making them extremely suitable for some implantable devices.

3.2 Long-term cycling stability

The lifetime of Li-ion micro-batteries is typically less than 1000 cycles, with charges retention higher than 98% after 1000 charge-discharge cycles in supercapacitors. Generally, these materials have lifetimes of over 100,000 cycles with excellent recharge rate capabilities.

3.3 Power performance

The power density is different in various active materials. For PPY, the capacity can reach 230 mF/cm^2, while for MnO_2, the capacity can reach as high as 570 F/g, and asymmetric micro-SC has a maximum volumetric energy density. The supercapacitors shown high-power density with many other advantages, the low-energy density limits their applications. The future trend of supercapacitors is still to overcome this drawback and find the active materials with more energy density. For energy transform devices, the unstable voltage supplied by energy harvesting limits their applications; thus, we will not compare their performance.

4. Technologies for fabricating supercapacitors

Various structures can be observed supercapacitors, including sandwich structures, interdigitated structures, MXene fiber structures, and core-shell structures. Sandwich structures generally consist of one, composed of two electrodes surrounding a separator at its center, with numerous technologies available to build electrodes, which can be further packaged into sandwich structures. Some supercapacitors, especially micro-supercapacitors, may not require separators, as the electrodes can be fixed without a connection between them. In this section, we list some methods for designing and fabricating electrodes and decorating active materials in sandwich and interdigitated structures.

4.1 Lithography

Lithography has been used to design and fabricate interdigitated circuit chips with an ultra-nanoscale size, making it extremely easy to fabricate micro-supercapacitors. Using lithography, the supercapacitor can integrate a circuit chip on the silicon wafer. These types of supercapacitors do not require separators, as they do not contain a short circuit with a fixed and hard substrate. Porous silicon and an interdigitated shape are generally used, as they exhibit a large area. Researchers have also designed and fabricated various flexible and transparent supercapacitors on special templates using lithography.

4.2 Electrospinning

Electrospinning can be used to fabricate a designed nanoscale space pose, making it suitable for the electrodes of supercapacitors. Long nanofibers have been produced into a thin film through electrospinning, and then metals were coated or deposited onto the surfaces of these pre-electrospinning films, where the diameter and thickness of these films could be controlled by the diameters of the nanofibers.

4.3 Inkjet printing

Dedicated inks consisted of metal nanoparticles, CNTs, and graphene, and after mixing with the gel, these inks could be injected and printed onto the template surface, forming an active material. They can usually be fabricated in the shapes of interdigitated electrodes to increase the active area.

4.4 Laser scribing

Various micro- and nanofabrications can be used to build supercapacitors. Direct laser writing also offers the potential to realize a lower cost and largely scalable fabrication [43]. For the preparation of active materials films, which are composed of CNT, metal oxides, and graphene, lasers have generally been used to fabricate interdigitated electrodes, with large capacitors.

4.5 Electrophoretic deposition, electrolytic deposition, physical/chemical vapor deposition, and spurting

Many methods are available to attach electrodes with active materials, such as electrophoretic deposition, electrolytic deposition, physical/chemical vapor deposition [44], and spurting. The substrate and collectors can be deposited by atmospheric application, where the active materials will be deposited on the conductive metal-organic frameworks. The same method has also shown to be suitable for use in conductive polymers, which are also one of the main active materials with high quality and stability.

4.6 Sacrificial template

In terms of geometries, key design considerations can be used to increase the surface areas of electrodes, where a sacrificial template is used to increase the area of the electrode. These sacrificial templates include AAO film, polystyrene colloidal particles film, and cube sugar. Template films can be built with sacrificed nanoparticles, which can be dissolved with the target reagent. Pre-curve PDMS has usually been used as the supporting film, while the AAO template has also shown to be a fantastic sacrifice film with a controllable shape, on the nanoscale [30].

4.7 Transfer printing technologies and screen printing

Transfer printing and screen printing are mature technologies that transfer the shape from the donor wafer to the receiving substrate. After printing, the inks will be released from the stamp to form the desired functional layouts on the receiver substrate. The inks for transfer printing include various organic materials and carbon materials, which can be used to fabricate the flexible and stretchable circuits and supercapacitors [45].

Over the discussion mentioned earlier, many technologies have already been used to design and fabricate the supercapacitors; however, the main works is still focusing on increasing the energy density of the active materials and shape designing for active materials. These new fabrications technologies are hopefully to increase the density of the supercapacitors with suitable active materials.

5. Applications of supercapacitors

As a power storage technology, there are still many challenges for the practical applications of supercapacitors, and the major drawback of supercapacitors is their low energy density. Significant efforts have been made to improve their performance. Unlike work on main power storage systems, supercapacitors have shown to be more important as battery complements, especially due to the many advantages of supercapacitors compared with lithium-ion batteries, such as a highly charge ability, long-term stability, and power density, as well as flexibility, transparency, microscale, and stretchability. Furthermore, the ions and electrolyte are both the key considerations at the electrode-electrolyte interface for various applications. The biocompatible electrolytes and ions are more suitable for solid-state and stretch ability supercapacitors using in wearable applications, while the highly diffusing rate of ions can be used in hybrid electric vehicles and solar cells with low impedance. The dynamic of the ion at the electrode-electrolyte interface endows electrochemical energy storage apparatuses. Finally, there are many applications, and we list some important samples, including portable electronics [43], solar cells, hybrid electric vehicles [46], and wearable devices [5, 47].

5.1 Vehicles

Vehicles are common devices, as fossil fuels are no longer sustainable and result in significant pollution. To achieve efficient energy management systems [6], combination of batteries and supercapacitors has been proposed, as the supercapacitor can absorb energy from braking and provide energy when powered on, and these electrochemical energy storage apparatuses are based on the highly diffusing rate at the electrode-electrolyte interface with high energy or power density, long lifetime, and high safety insurance. This energy management system has been used in many electronic buses, and the same saving power technology has been used in elevators and trains.

5.2 Integration of solar cells and supercapacitors

Solar energy [48] is the main energy source for all plants and humans, with a major utilization of solar energy relying on photovoltaic technology; however, this process is unstable because solar radiation is intermittent and unstable [49], which can destroy the lifetime of solar cells and devices. Advanced approaches have been proposed using electric energy storage systems, where the integration of supercapacitors and solar cells [50] consisted of three parts, namely, dye-sensitized solar cells, perovskite solar cells, and organic solar cells. Overall, supercapacitors have shown the potential to be next-generation power sources [51], especially for providing power supply over a short amount of time and working as an energy buffer and integrating with other power storage systems.

5.3 Supercapacitors for wearable devices

Wearable systems offer a considerable amount of health information, such as heart rate, electrocardiogram, and activity data, and these devices have rapidly gained market approval. Driven by the rapid growth of portable and wearable electronics, significant research attention has been focused on the development of energy storage devices. Traditionally, energy is stored in Li-ion micro-batteries; however, their lifetime usually imposes the need for costly and impracticable maintenance, and the integrated devices and immune reactions restrict periodic part replacements. Furthermore, the hard shell and toxicity of the electrolyte in coin cells will be harmful

DOI: http://dx.doi.org/10.5772/intechopen.107419

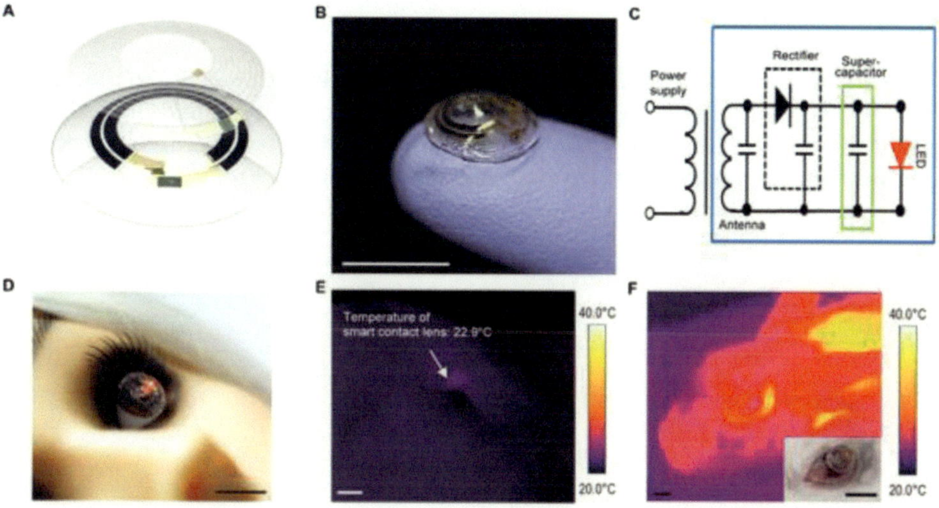

Figure 3.
Smart soft contact lens with wireless charge supercapacitor: (a) illustration of integrated contact lens, (b) photograph of the fully integrated device, scale bar 1 cm, (c) circuit diagram of supercapacitor recharge, (d) photograph of contact lens on an eye of a mannequin, scale bar 1 cm, (e) infrared radiation image of this contact lens on an eye of a mannequin, scale bar 1 cm, and (f) infrared radiation image and photograph (inset) during the discharging state on the eye of a live rabbit eye, scale bar 1 cm [52].

to organisms. The need for power for wearable electronics has motivated the development of suitable replacements. Supercapacitors are promising candidates for portable and wearable devices [37], as they should be light in weight, compact in size, and high in energy density and have a lifetime of over 100,000 cycles with excellent recharge rate capabilities. Furthermore, the biocompatible electrolyte and mild reactions at the electrodes-electrolyte interface shows highly potential being the batteries complements and new generation power storages, including remote sensors [23], implantable biosensors [52], and nanorobots. As can be seen, **Figure 3** is a sample supercapacitor application with integrated soft, smart contact lens with wireless charge supercapacitors. In this system, the supercapacitor worked as a complementary storage system for the lens, and it could charge with an antenna from the power supply. Moreover, supercapacitors may be completely biodegradable and bioabsorbable, which have never been achieved in other power storage systems [42].

6. Conclusions

Supercapacitors are devices for energy storage systems, which shown great potential as important complements to batteries. These devices consist of collectors, electrodes, active materials, separators, and electrolytes. We introduced the principle and components of supercapacitors, and then we compared the advantages of supercapacitors with other energy storage systems. In the third part, the approaches for supercapacitor fabrication were described. Finally, as complementary storage for batteries, supercapacitors are becoming increasingly important in certain applications. The applications of supercapacitors for electronic devices were described, indicating that supercapacitors may be more suitable for use in wearable and portable devices.

Acknowledgements

The authors acknowledge start-up funds from Westlake University to the Center of Excellence in Biomedical Research on Advanced Integrated-On-Chips Neurotechnologies (CenBRAIN Neurotech) for supporting this research project. This work was funded in part by the Zhejiang Key R&D Program Project No. 2021C03002 and in part by the Zhejiang Leading Innovative and Entrepreneur Team Introduction Program No. 2020R01005.

Conflict of interest

The authors declare no conflict of interest.

Author details

Yi Su and Mohamad Sawan*
CenBRAIN Neurotech Center of Excellence, Westlake University, Hangzhou, Zhejiang, China

*Address all correspondence to: sawan@westlake.edu.cn

IntechOpen

© 2022 The Author(s). Licensee IntechOpen. This chapter is distributed under the terms of the Creative Commons Attribution License (http://creativecommons.org/licenses/by/3.0), which permits unrestricted use, distribution, and reproduction in any medium, provided the original work is properly cited.

References

[1] Chandra A. Proceedings of the National Academy of Sciences. India Section A: Physical Sciences. 2012;**82**:79-90

[2] Salanne M, Rotesnberg B, Naoi K, Kaneko K, Taberna PL, Grey CP, et al. Nature Energy. 2016;**1**(6)

[3] Pan Z, Yang J, Zhang Q, Liu M, Hu Y, Kou Z, et al. Advanced Energy Materials. 2019;**9**:1802753

[4] Rapisarda M, Marken F, Meo M. Communications Chemistry. 2021;**4**(1)

[5] Yan J, Li S, Lan B, Wu Y, Lee PS. Advanced Functional Materials. 2019;**30**(2)

[6] Sampietro J, Puig V, Costa-Castelló R. Energies. 2019;**12**(5):925

[7] Zhang R, Ding J, Liu C, Yang E-H. ACS Applied Energy Materials. 2018;**1**:2048-2055

[8] Liu T, Yan R, Huang H, Pan L, Cao X, deMello A, et al. Advanced Functional Materials. 2020;**301**(5):2048-2055

[9] Qiu T, Luo B, Giersig M, Akinoglu EM, Hao L, Wang X, et al. Small. 2014;**10**(20):4136-4141

[10] Kiruthika S, Sow C, Kulkarni GU. Small. 2017;**13**(40):1701906

[11] Zhang C, Li H, Huang A, Zhang Q, Rui K, Lin H, et al. Small. 2019;**15**(18): e1805493

[12] Lee Y, Chae S, Park H, Kim J, Jeong S-H. Chemical Engineering Journal. 2020;**382**:122798

[13] Xiao Y, Liu Y, Liu F, Han P, Qin G. Journal of Alloys and Compounds. 2020;**813**:152089

[14] Wu S-R, Liu J-B, Wang H, Yan H. Ionics. 2018;**25**(1):1-8

[15] Yang Y, Zhang J, Zhang H, Hou Y, Guo J. Journal of Porous Materials. 2020;**27**(5):1309-1317

[16] Gilshtein E, Flox C, Ali FSM, Mehrabimatin B, Fedorov FS, Lin S, et al. Journal of Energy Storage. 2020;**30**: 101505

[17] Chen J, Wang Y, Cao J, Liao L, Liu Y, Zhou Y, et al. Electrochimica Acta. 2020;**361**:137036

[18] Deng Q, Abbas SC, Li Z, Lv J, Ma X, Cao S, et al. Journal of Colloid and Interface Science. 2020;**574**:33-42

[19] Shieh J-Y, Tsai S-Y, Li B-Y, Yu HH. Materials Chemistry and Physics. 2017;**195**:114-122

[20] Kulandaivalu S, Sulaiman Y. Journal of Power Sources. 2019;**419**:181-191

[21] Esfahani MZ, Khosravi M. Journal of Power Sources. 2020;**462**:228166

[22] Zeng J, Dong L, Sha W, Wei L, Guo X. Chemical Engineering Journal. 2020;**383**:123098

[23] Rajendran V, Mohan AMV, Jayaraman M, Nakagawa T. Nano Energy. 2019;**65**:104055

[24] Wu Y, Hu H, Yuan C, Song J, Wu M. Nano Energy. 2020;**74**:104812

[25] Li X, Hao J, Liu R, He H, Wang Y, Liang G, et al. Energy Storage Materials. 2020;**33**:62-70

[26] El-Kady MF, Ihns M, Li M, Hwang JY, Mousavi MF, Chaney L, et al. Proceedings of the National Academy of

Sciences of the United States of America. 2015;**112**(14):4233-4238

[27] Park H, Kim JW, Hong SY, Lee G, Kim DS, Oh JH, et al. Advanced Functional Materials. 2018;**28**(33):1707013

[28] Chen S, Shi B, He W, Wu X, Zhang X, Zhu Y, et al. Advanced Functional Materials. 2019;**29**(48):1906618

[29] Jian Z, Yang N, Vogel M, Leith S, Schulte A, Schönherr H, et al. Advanced Energy Materials. 2020;**10**(44):2002202

[30] Lei Z, Liu L, Zhao H, Liang F, Chang S, Li L, et al. Nature Communications. 2020;**11**(11):299

[31] Kou L, Huang T, Zheng B, Han Y, Zhao X, Gopalsamy K, et al. Nature Communications. 2014;**5**:3754

[32] Wu G, Sun S, Zhu X, Ma Z, Zhang Y, Bao N. Angewandte Chemie (International Ed. in English). 2022;**61**(8):e202115559

[33] Meng J, Wu G, Wu X, Cheng H, Xu Z, Chen S. Advanced Science. 2020;7(1):1901931

[34] Tian J, Cui N, Chen P, Guo K, Chen X. Journal of Materials Chemistry A. 2021;**9**(36):20635-20644

[35] Su F, Wu Z-S. Journal of Energy Chemistry. 2021;**53**:354-357

[36] Nguyen T, Montemor MF. Advanced Science. 2019;**6**(9):1801797

[37] Su Y, Li N, Wang L, Lin R, Zheng Y, Rong G, et al. Advanced Materials Technologies. 2021;7(1):2100608

[38] Yu H, Tang Q, Wu J, Lin Y, Fan L, Huang M, et al. Journal of Power Sources. 2012;**206**:463-468

[39] Maitra A, Karan SK, Paria S, Das AK, Bera R, Halder L, et al. Nano Energy. 2017;**40**:633-645

[40] Tan R, Wang A, Malpass-Evans R, Williams R, Zhao EW, Liu T, et al. Nature Materials. 2020;**19**(2):195-202

[41] Abdel Maksoud MIA, Fahim RA, Shalan AE, Abd Elkodous M, Olojede SO, Osman AI, et al. Environmental Chemistry Letters. 2020;**19**(1):375-439

[42] Sheng HW, Zhou JJ, Li B, He YH, Zhang XT, Liang J, et al. Science Advances. 2021;**7**(2):eabe3097

[43] Gao W, Singh N, Song L, Liu Z, Reddy AL, Ci L, et al. Nature Nanotechnology. 2011;**6**(8):496-500

[44] Li X, Sun P, Fan L, Zhu M, Wang K, Zhong M, et al. Scientific Reports. 2012;**2**:395

[45] Linghu C, Zhang S, Wang C, Song J. npj Flexible Electronics. 2018;**2**(1)

[46] Olabi AG, Abbas Q, Al Makky A, Abdelkareem MA. Energy. 2022;**248**:123617

[47] Keum K, Kim JW, Hong SY, Son JG, Lee SS, Ha JS. Advanced Materials. 2020;**32**(51):2002180

[48] Plebankiewicz I, Bogdanowicz KA, Iwan A. Energies. 2020;**13**(15):3867

[49] Tian Z, Tong X, Sheng G, Shao Y, Yu L, Tung V, et al. Nature Communications. 2019;**10**:4913

[50] Nguyen DCT, Shin J, Kim S-K, Lee S-H. ACS Applied Energy Materials. 2021;**4**(12):14014-14021

[51] Zeng Q, Lai Y, Jiang L, Liu F, Hao X, Wang L, et al. Advanced Energy Materials. 2020;**10**(14):1903930

[52] Park J, Ahn DB, Kim J, Cha E, Bae BS, Lee SY, et al. Science Advances. 2019;**5**(12):eaay0764